BRAHIM EL GHMARI
ABDELLAH ECH-CHAHAD
HANANE FARAH

Biosynthesis, characterization and application of nanoparticles

BRAHIM EL GHMARI
ABDELLAH ECH-CHAHAD
HANANE FARAH

Biosynthesis, characterization and application of nanoparticles

ScienciaScripts

Imprint

Any brand names and product names mentioned in this book are subject to trademark, brand or patent protection and are trademarks or registered trademarks of their respective holders. The use of brand names, product names, common names, trade names, product descriptions etc. even without a particular marking in this work is in no way to be construed to mean that such names may be regarded as unrestricted in respect of trademark and brand protection legislation and could thus be used by anyone.

Cover image: www.ingimage.com

This book is a translation from the original published under ISBN 978-620-6-72506-0.

Publisher:
Sciencia Scripts
is a trademark of
Dodo Books Indian Ocean Ltd. and OmniScriptum S.R.L publishing group

120 High Road, East Finchley, London, N2 9ED, United Kingdom
Str. Armeneasca 28/1, office 1, Chisinau MD-2012, Republic of Moldova, Europe
Printed at: see last page
ISBN: 978-620-6-16613-9

Contents

Introduction generate

MétaLLque oxide nanoparticles are a class of matërials with particle sizes between 1 and 100 nanomëtres. These nanoparticles have unique physical and chemical propriëtës that differ from those of bulk matërials due to their ëκyë surface-to-volume ratio, quantum confinement effects and surface plasmon resonance. As a result, mëtalic oxide have foundë applications in various fields, including catalysis, electronics, energy, biomëdicine and environmental remediation[1].

One of the main advantages of mëtalic oxide is their high reactiv^, which makes them useful for catalysis. They can be used as catalysts in various chemical reactions, such as the reduction of nitric oxide, the oxidation of carbon monoxide and the hydrogenation of organic compounds. The ëκyë surface-to-volume ratio of mëtalic oxide provides a large number of active sites for catalytic reactions, resulting in more ëlevës[2] reaction rates and yields.

In the field of electronicsmëtalic oxide nanoparticles are used as basic ëlëments for the fabrication of nanoscale devices. These nanoparticles can be synthëtisëed with spëcific shapes and sizes, which allows their ëtës electronic properties to be controlled. Mëtalic oxide also exhibit unique optical properties, such as plasmon resonance, making them suitable for applications in detection, imaging and optoelectronics[3].

In the energy field, mëtalic oxide are used as electrode materials in lithium-ion batteries, which are widely used in portable ëlectronic devices. The surface-to-volume ratio ëκyë of mëtalic oxide provides a large number of sites for lithium-ion intercalation, resulting in higher energy densities ëlevëes and longer battery life[4].

In the biomëdical field, mëtalic oxide have found applications in drug delivery, imaging and cancer therapeutics. They can be functionalised with targeting ligands to selectively deliver drugs to specific cells or tissues[5]. Metal oxide can also be used as contrast agents for magnetic resonance imaging (MRI) due to their magnetic properties[6]. In cancer therapy, metal oxide can be used for hyperthermia treatment, which involves heating cancer cells using an external magnetic field[7].

In the field of environmental remediation, mëtaLLque oxide nanoparticles have found applications in the elimination of water and air pollutants. They can be used as catalysts to oxidise pollutants or as adsorbents to remove them from the environment[8].

Overall, mëtalic oxide are a versatile class of matërials with many applications in various fields. The unique physical and chemical propriëtës of these nanoparticles make them attractive for a wide ëvent applications, and their use is expected to continue to dëvelop in the future[7], [9], [10].

The biosynthesis of nanoparticles, including silver and nickel oxide , is a fascinating area of research that has attracted a great deal of attention due to its potential applications in various fields [1], [11], [12]. The motivation for the biosynthesis of these nanoparticles can be attributed to several factors:

❖ Green and sustainable synthesis: Traditional methods of synthesising nanoparticles often involve the use of harsh chemicals and high temperatures, which can be harmful to the environment and consume a lot of energy. Biosynthesis offers a greener and more sustainable alternative by using biological agents such as micro-organisms, plants or enzymes to carry out the synthesis under milder conditions[7].

❖ Biocompatibility: Biosynthesised nanoparticles tend to be more biocompatible than those produced by chemical methods. This is particularly important for medical applications, such as drug delivery and imaging, where nanoparticles must interact safely with living organisms[5].

❖ Size and shape control: Biosynthesis techniques provide better control over the size, shape and morphology of nanoparticles. This control is vital because these factors can significantly influence the properties and applications of nanoparticles.

❖ Low cost: in many cases, the biological agents used for biosynthesis can be readily available and inexpensive, helping to reduce the overall production cost of nanoparticles.

❖ Biological models: Certain organisms, such as bacteria and plants, can serve as natural models for the synthesis of nanoparticles. Nanoparticles can be formed on or in these models, giving them unique propriëtës and potential applications in ouutulyulu, electronics, etc.[3], [13].

❖ Атё^^^ of catalytic activity: mëtalic nanoparticles, such as silver and nickel oxide nanoparticles, often exhibit increased catalytic activ^ due to their surface-to-volume ratio ëкуë. These nanoparticles can be utilisëes in various catalytic processes, including environmental dëpollution and industrial chemical synthesis[13], [14].

❖ Antibacterial properties: Silver nanoparticles, for example, are known for their strong antibacterial properties. Biosynthëtisëed silver can be used in the medical and healthcare field for wound healing, stërilisation and the control of antibiotic-resistant bacteria.

❖ Applications in the field of sensors: Biosynthëtisëed nanoparticles can be used as ëlë detection elements in various devices. Their unique ëtëlectronic and optical properties enable them to create sensitive and seëlective sensors for the dëtection of gases, chemicals and biomoUcules.

❖ Applications ënergëtiques : Silver and nickel oxide have potential applications in the Këз a l^nergie fields. Silver nanoparticles can be used in solar cells, catalysis for fuel cells, etc. Nickel oxide can find applications in supercapacitors, sensors and energy storage devices.

❖ Advances in nanotechnology: Nanoparticle biosynthesis is helping to advance the field of nanotechnology, enabling the dëvelopment of innovative materials and devices with improved properties and performance.

So, our motivation for the biosynthesis silver and nickel oxide nanoparticles comes from the dësir de_svntlK'tiser des nanoparticles ayant des propriëtës controls et une biocompatibil^ amëliorëe avec un impact environnemental reduit qui peuventre utilisées dans divers domaines tels que la mëdecine, la catalyse, l^nergie et l^lectronique, etc....

CHAPTER I

Ethnobotanical study of the plant H. Hirsuta

I. Introduction

H. hirsuta, often referred to as the hernia grass, is a plant that has found its place not only in arid regions, but also in the world of botany and ecology. Its deep roots, lance-shaped leaves and creeping habit make it a species adapted to the harshest environments. However, it's not just its adaptation to harsh conditions that makes it special. *H. hirsuta* also plays a crucial role in preserving soil and biodiversity in these arid environments. Its roots act as a real anchor for the soil, erosion and helping to maintain the integrity of fragile ecosystems. In addition to this, this plant has fascinating botanical properties, such as its ability to reproduce autonomously by self-fertilisation, making it an important subject of study for ecology researchers. *H. hirsuta* is therefore an extraordinary example of the way plants have evolved to withstand extreme environmental conditions while playing a key role in nature conservation.

II. Taxonomic classification of *H. hirsuta*

H. hirsuta is a species of flowering plant in the family Caryophyllaceae. Here is a summary of its taxonomic classification [15]:
- **Kingdom**: Plantae
- **Subregnum**: Tracheobionta
- **Superdivision**: Spermatophyta
- **Division**: Magnoliophyta
- **Class**: Magnoliopsida
- **Subclass**: Caryophyllidae
- **Order** : Caryophyllales
- **Family**: Caryophyllaceae
- **Genre**: Herniaria
- **Species**: *H. hirsuta*

There are also some variations in the taxonomic classification of *H. hirsuta*, including subspecies and varieties. For example, *H. hirsuta var. hirsuta* and *H. hirsuta ssp. cinerea* are recognised by the Integrated Taxonomic Information System (ITIS) as variëtës and subspecies of *H. hirsuta ,* respectively. However, the taxonomic status of these variations may vary depending on the source.

III. Nomenclature of the plant *H. hirsuta*

Here are some common names for the *H. hirsuta* plant [16].
- **Arab**: Harras lahjar, Hashishat Al Fatik, Marda, Noman amrad, Dizama;
- **English**: Hairy rupturewort,
- **French**: Herniaire velue; casse piërres.
- **German**: Behaartes Bruchkraut, Behaartes Bruchkraut;
- **Sweden**: Luddknytling.

IV. Morphology of *the H. Hirsuta*

The plant *H. hirsuta* is an herbaceous species belonging to the family **Caryophyllaceae**.

Here is a dëtaillëe description of its morphology [17], [18]:

■ **Root:** *H. hirsuta* possëde a shallow root system, with shallow roots that extend horizontally into the soil.

■ **Stem:** The stems of hërissëe are slender, creeping and generally prostrate on the ground. They can reach a length of 10 to 30 cm. The stems are ramii'ie from the base and can form dense tufts.

■ **Leaves:** The leaves of this plant are opposite and arranged along the stems. They are sessile, meaning that they do not have a distinct pëtiole, and they are gënëralement oblong a spatu^es. The leaves are covered with short, soft hairs, giving them a velvety texture.

■ **Flowers:** The flowers of the hernia hërissëe are small and inconspicuous. They are generally pale green to yellowish white. They are grouped into small, dense inflorescences forming clusters or axillary cymes. Each flower has five sëpals fused at the base and five pëtals reduced to the shape of scales. The plant in monoecious, meaning that it bears both male and female flowers on the same individual.

■ **Fruit and seeds:** The plant produces small capsule-shaped fruits which contain seeds. The capsules are usually dëpourvues teeth or protubërances and may contain 5 to 10 seeds.

In rësumë, *H. hirsuta* is a small creeping plant with slender stems, opposite velvety leaves, inconspicuous green to white flowers and seed capsules. Its leaves are small, light green and up to a centimetre long and border the stems (Figure I.1).

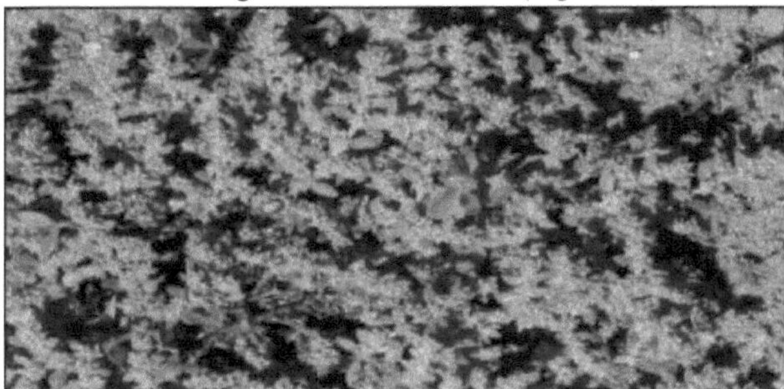

Figure I. 1: Photo of the *H. hirsuta* plant.

V. Geographical distribution of the plant *H. hirsuta*

V.1 Worldwide

H. hirsuta has a wide geographical distribution in various parts of the world, with a presence in Europe, North Africa and North America. It exists in disturbed environments, particularly uncultivated land and predominantly sandy soils, from sea level to the lower mountain belt and, in all cases, up to 1,600 mëtres above sea level.

Figure I.2 [19] shows the distribution of this plant in different countries around the world:

- In Africa: Ethiopia, Algeria, Egypt, Morocco.
- In Asia: Kuwait, Armenia, Azerbaijan, Russian Federation, Kyrgyzstan, Tajikistan, Turkmenistan, Uzbekistan, Iran, Iraq, Palestine, Jordan, Lebanon, Syria, Turkey.
- In Europe: Austria, Belgium, Germany, Hungary, Slovakia, Switzerland, Albania, Bulgaria, Croatia, Greece, Italy, Macedonia, Romania, Slovenia, France, Portugal, Spain.

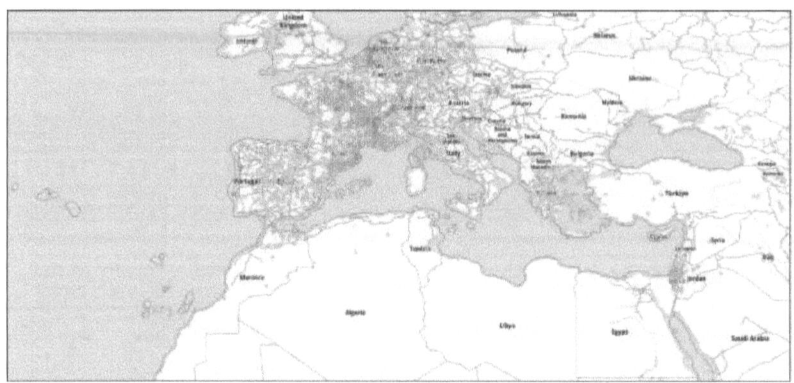

Figure I. 2: Worldwide distribution of the plant Herniaria hirsuta.

V. 2. In Morocco

Although specific studies on the distribution of *H. hirsuta* in the different regions of Morocco are limited, the information available suggests that the plant can be found in different parts of the country [20], [21]. According to ethnobotanical studies, *H. hirsuta* is present in the Rabat, Oujda and Fes-Boulemane regions of Morocco [22].

It is listed as present in Morocco on the INPN (Inventaire National du Patrimoine Naturel) website, which provides information on the distribution of species in France and overseas territories [23].

VI. Chemical composition

The chemical composition of *H. hirsuta* can vary depending on a number of factors, including geographical location, harvesting period, climatic factors, edaphic factors and the extraction used. Previous studies have shown that aqueous extracts of *H. hirsuta* are rich in phytochemicals such as saponins, flavonoids, coumarins, triterpenic acids and sterols. A study published in the journal "Natural Product Research" gives an example of the approximate chemical composition *of* H. *hirsuta* presented in *Table I.1* [24]:

Table I. 1: Example of the chemical composition of *H. hirsuta*.

Families	Mass quantity %o	Chemical compounds	Mass quantity %o
Flavonoides	1,08 %.	Apigenin	0.52%
		Luteolin	0.34%
		Quercetin	0.18%
Triterpenoids	0.32%	Ursolic acid	0.17%
		Oleanolic acid	0.15%
Sterols	0.11%	P-Sitosterol	0.08%
		Stigmasterol	0.03%
Alkaloids	0.03%	Herniarine	0.02%
		Herniaridine	0.01%

VII. Pharmacological effects

H. hirsuta is used in traditional Moroccan medicine for the treatment of arterial hypertension, urinary and biliary lithiasis [25]. According to previous studies, this plant possesses numerous pharmacological effects such :

- **Anti-urolithiasis**: Numerous studies have shown that aqueous extracts of *H. hirsuta* and *H. glabra* are effective in the treatment and prevention of urothiasis in rats [26], [27]. Meiouet et al (2010) have shown that *H. hirsuta* extract is capable of dissolving cystine lithiasis through the formation of cystine-saponin or cystine-flavonoid complexes [28].

- **Diuretic**: Mabrouki et al (2021) have shown that ethanolic extracts of *H. glabra* have diuretic activity. This activity is linked to the saponins present in these extracts[29].

- **Antimicrobial**: *H. glabra*, a related species, has been shown to have antimicrobial effects [23],[30].

- **Antioxidant** : According to the study by Kolodziejczyk-Czepas et al (2021) *H. glabra* has been shown to have antioxidant effects[31].

VIII. Conclusion

I. *hirsuta* is a modest-growing annual plant native central and southern Europe, but introduced to various parts of the world. Its traditional use in medicine dates back several centuries, mainly for its properties such as anti-urolithiasis, diuretic, antimicrobial and antioxidant. In addition, traditional Moroccan medicine has also used it to treat conditions such as biliary dyskinesia, urolithiasis and as a diuretic.

Despite this traditional use, there have been few in-depth studies to scientifically assess the various applications and benefits of this plant.

CHAPTER II

State of the art on nanoparticles

II. Introduction

The prefix "nano", of Greek origin, means "dwarf," and it is ий^ ё to denote the domain of the infinitely small [46]. The cara^ristic scale varies approximately from 1 to 100 nanometres (nm). Indeed, we are talking here about extremely small materials at the nanometre scale. This is equivalent to 1/100 the width a DNA molecule or 1/50000 the thickness of a human hair [47]. The world of nanosciences and nanotechnologies, known as the "nanoworld", is attracting unprecedented interest from government institutions, research centres, public and private universities, as well as certain companies, which are studying these fundamental resources on a global scale. Nanoscience and nanotechnology are areas of materials science research that focus on the manipulation and use of matter at the nanomëtric scale. These areas of research have become increasingly important and popular in recent decades due to their revolutionary potential for medicine, electronics, energy, the environment and many other fields. The technological development of nanoscience involves the use of nanomaterials and objects to create structures, devices and systems. These nanoscale materials exhibit specific quantum, surface, physical and chemical properties that differ from the corresponding bulk materials.

However, it is important to note that this nanoscience technology is still in its infancy, and it is difficult to estimate and track future scientific and technological advances with any certainty [31].

III. Historical aspects

Nanoparticles have been used since Roman times for the dëcorative colouring of glass. However, at that time, the very concept of nanoparticles was unknown [32].

Surprisingly, nanotechnology was widely used in the Attica region as early as the 6th century BC [33]. During the Archaic and Classical periods (around 620-300 BC), the production of dëcorës vases in this region reached remarkable artistic heights thanks to a highly innovative manufacturing technique. As part of this technique, spinel nanoparticles were formed inside a glassy layer a few microns thick [34].

Celtic red ëтаих, dating from 400 *to* 100 BC, contained copper and cuprous oxide nanoparticles, while the majority of red tesserae usedëes in Roman mosaics ёt were made from glass containing copper nanocrystals [35], [36].

Between 1066 and 1485 AD, ëmesh cëramics with striking optical effects appeared, achieving their beautyë through the use of mëtalic nanoparticles [37].

In ancient Indian and Chinese mëdecines, soluble gold was used therapeutically, containing gold nanoparticles mëlangëed to larger particles [38]. However, ancient civilisations did not understand the unique propriëtës and potential of their preparations as we now understand them [39].

A revolution in nanotechnology began with Faraday's work on colloidal gold in 1857, which interest in metallic nanoparticles [40]. In 1959, a confërence, physicist Richard Feynman dëclaimed, "The principles of physics, as far as we can tell, are not opposed to the possibility of manipulating things atom by atom." This statement paved the way for the scientific community to explore the universe of the infinitely small [33].

The term "nanotechnology" was first coined in 1974 by Norio Tanigushi, although it was used earlier by Eric Drexler in his 1986 book "Engines of Creation: The Coming Era of Nanotechnology" [41]. The history of nanotechnology has seen a fascinating evolution, marked by some key discoveries:

> **1981:** Development of the scanning tunneling electron microscope, which allows individual atoms to be manipulated.

> **1985:** Discovery of fullerene, a circular structure of 60 carbon atoms (C60).

> **1992:** Identification of carbon nanotubes, which are stronger than steel and used for applications such as drug delivery and energy storage and transmission.

> **1993:** Discovery of quantum dots, which have interesting optical properties.

> **2000:** Manufacture of passive nanoparticles used in a variety of applications, including nanofuel cells and cosmetics

> **2005:** Development of active nanoparticles for drug targets and other adaptive structures.

These ët milestones have 1raeë the way explore the infinitely small and opened up exciting prospects for the future of nanotechnology.

IV.. Definitions

IV.1. 1. Nanosciences

Nanoscience is an interdisciplinary field that focuses on the design and creation of functional systèmes at a molecular or atomic ëchelle. It is consideredërë to be a major branch of science, offering the ability^ to understand, manipulate and exploit matter at the atomic scale [42]. Fundamentally, nanoscience aims to explore and exploit the unique propertiesëtës and behaviours of matter at extremely small levels. Given that everything that constitutes matter is made up atoms and molecules, nanoscience potentially extends to all areas of science and technology, offering incredibly vast prospects for innovative applications and revolutionary technological advances.

IV.2. 2. Nanotechnology

Nanotechnologies represent the major technological innovation of the 21st century and are considered by many specialists as the beginning a new industrial era, that of the technologies of the small [43]. The branch of nanotechnology focuses particularly on the study of processes that take place at the molecular level and at the nanomëtric scale. The main aim of this discipline is to synthesise new nanoparticles of various sizes and morphologies. However, nanotechnologies are not limited to simple miniaturisation; they are often characterised by the integration of new emerging laws of behaviour that dominate the operation of the objects produced[44].

Nanotechnology is an essential area of research in the context of advanced research, with potential applications in various sectors. Its remarkable growth is opening up new prospects at both the applied and fundamental levels, particularly in the synthesis of materials at the nanomëtric scale and in the understanding of their fascinating optical and physicochemical properties [45].

IV.3. 3. Nanomaterials

Nanomaterials are materials composed in whole or in part of nano-objects, which have specific properties due to their size. These nanomatërials can be classified into three distinct catëgories [1]:

❖ Materials that incorporate nano-objects into an organic or moral matrix to provide new functionalities or modify mechanical, optical, magnetic or thermal properties.

❖ Materials whose surface is nanostructured in order to conferëtës properties such as abrasion resistance or hydrophilicity, or to provide new functionalities such as adherence and скп'сЧё.

❖ Materials whose intrinsic structure is nanostructured in volume, thus giving them particular physical properties.

III.4. Nanoparticles

Nanoparticles are structured assemblies of hundreds to thousands atoms forming objects with at least one dimension between 1 and 100 nm. These nano-objects are located at the interface of the macroscopic (**mass** matëriau) and molecular (or atomic) scales [46]. By consëquent, this dëfinition excludes objects whose smallest dimensions are between 100 and 1000 nm. Although these particles are nanomëtric in size, they are called submicron particles. Compared with natural organic structures, nanoparticles are mainly in the size range corresponding to protëines (Figure II.1) [47].

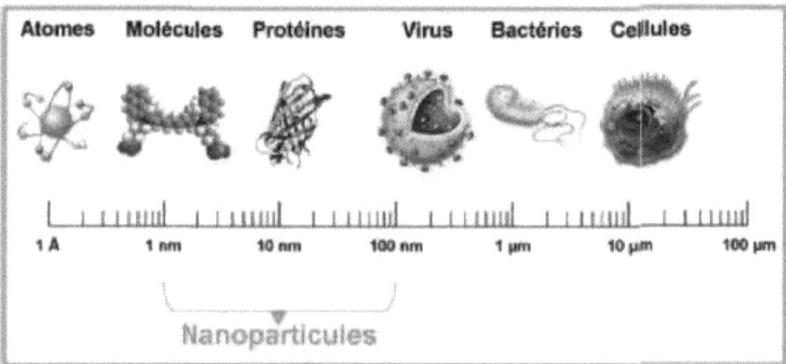

Figure II. 1: Size range of nanoparticles compared to the main chemical and biological chemical and biological structures.

Most micron-scale materials have the same physical properties as bulk materials.

On the other hand, *a* ГёсЬеПе папотёй^ ие, they can have very different physical propriëtës to bulk matërials. These diffërences give nanoparticles particular propriëtës [48]. As such, they are consideredërës as building blocks of new дё^ тО:^ in various fields such as chemistry, physics, lklectronics, mëcanics and biotechnology [44], [49].

IV. Classification of nanoparticles

Nanoparticles can be classified in different ways according to various criteria such :
- Their chemical composition,
- Their size or shape,
- Their structure and origin.

IV.1 Classification of nanoparticles according to their size

Nanomaterials can be classifiedës according to different criteria, such as their dimensions, chemical composition, spëcific propriëtës or applications. Here are the classifications дё^ гакэ according to the dimensions of nanomatërials [50]:

10

IV.1.1. Nanoparticles of dimension 0 (0D)

Nanomatërials of dimension 0, ёдalement appeks nanomatëriaux 0D, are nanomëtric-scale matërials that have no ëtended dimensions in space. In other words, they are matërials that are coning to a point or structure without significant spatial extension. The term "0-dimensional nanomatërials" may seem counterintuitive, as nanomatërials are supposed to be small particles or structures in all dimensions (length, width, height). However, in the context of nanomatërials, the notion of dimensionalik can be différent. Common examples of 0D nanomatërials include:

4- **Nanoparticles:** individual particles of any size at the nanomëtric scale ^^ rally a few nanometres or less). They can be composedës of various materials, such as nanoparticles of gold, silver, silicon dioxide, etc. Nanoparticles often exhibit unique propriëtës due to their small size.

4- **Quantum dots:** These are semiconductor nanoparticles that possess special optical and ëlectronic properties due to their quantum confinement. They are used in applications such as advanced displays and biomedical imaging.

4- **Atomic clusters: an** aggregate of a small number atoms (usually a few dozen) forming a coherent whole, but without the periodic structure of crystals. Metal clusters have attracted a great deal of interest because of their catalytic and optical properties.

4- **Individual molecules:** Some molecules can be considered as zero-dimensional nanomaterials because of their nanometric size and their unique properties at this scale.

0-dimensional **(0D)** nanomaterials play a vital role in nanotechnology and have a wide range applications in electronics, medicine, energy and advanced materials. Because of their size at the nanometre scale, they often exhibit different properties to their larger counterparts, opening the door to new applications and technologies.

IV.1.2. Nanoparticles of dimension 1 (1D)

Also known as one-dimensional nanomaterials, are materials on the nanometric scale in which one dimension extends in one direction, but the other two dimensions are restricted or very limited. In other words, they are materials with a one-dimensional structure. Common examples of 1D nanomaterials include :

4- **Nanowires:** These are elongated materials with nanometric diameters, but which can be longer in one direction. Nanowires can be made of different materials such as silicon, carbon, zinc, etc. They are used in variety of applications, including electronic devices, sensors and solar cells.

4- **Nanotubes:** Nanotubes are hollow structures formed from coiled sheets of graphene. They can be monolayered or multilayered, with nanometric diameters and variable lengths.

4- **Nanofibres:** Nanofibres are fine, one-dimensional fibres with a diameter of nanometres. They can be made from a variety of materials, including polymers, metal oxides and proteins. Nanofibres are used in areas such as filtration, composite materials and tissue regeneration.

4- **Nanoribbons:** These are elongated structures of atomic thickness that can be fabricated from materials such as graphene or molybdenum disulphide. Nanoribbons have interesting electronic properties and could be used in ultra-thin electronic devices.

One-dimensional nanomaterials exhibit unique propriëtës due to their one-dimensional structure, which can make them useful in various nanoëlectronic, optical and biomëdical applications. Precise control of their size, shape and propriëtës is an active area of research in nanotechnology.

IV.1.3. Nanoparticles of dimension 2 (2D)

Two-dimensional nanomaterials, also known as 2D nanomaterials, are nanoscale materials with two extended dimensions and one very fine or finite dimension. In other words, they are materials with a two-dimensional structure formed by atomic layers. The best-known examples of 2D nanomatërials are 2D matërials such :

4- Graphene: Graphëne is a single layer of carbon atoms arranged in a two-dimensional hexagonal lattice. It is very strong, flexible and excellent ëlectronic properties, making it useful in a varietyëtë of applications, including electronics, catalysis and composites.

4- Molybdenum disulphide (MoS2): molybdenum disulphide is a two-dimensional material consisting a layer of molybdenum and two layers of sulphur. It also excellent ëtës ëlectronic properties and is usedë in opto-ëlectronic devices and sensors.

4- Hexagonal boron nitride (h-BN): Hexagonal boron nitride is a two-dimensional structure similar to graphëne, but with boron and nitrogen atoms. It is widely usedë as a thermal insulator and ëк^ нд^ in ëlectronic ëequipment.

4- Tungsten diseleniide (WSe2): Tungsten disëlëniide is another 2D matërial with interesting optical and ëlectronic propriëtës. Its applications in ëlectronic and photonic devices are ëtudiëes.

Two-dimensional nanomaterials have attracted a great deal of т1ёrё1: in nanotechnology research due to their unique properties at the atomic scale and their potential to create new technologies and applications. The propriëtës of these materials are strongly influenced by their size, shape and number of layers, so they can be adapted to meet specific needs in various fields of Гтдёшеnе and materials science.

IV.1.4. Nanoparticles of dimension 3 (3D)

Three-dimensional nanomatërials, ë also known as 3D nanomatërials, are nanoscale matërials with dimensions ëtended in all three directions of space. Unlike 0D, 1D and 2D nanomatërials, which have one- or two-dimensional structures, 3D nanomatërials have three-dimensional structures. 3D nanomatërials can be solid nanostructures with nanomëtrical dimensions in all three directions, or nanostructural matërials with spëcific nanomëtrical organisation, such as nanocrystals, nanocomposites, nanopowders, etc. Here are some examples of 3D nanomatërials:

4- Three-dimensional nanoparticles: These nanoparticles have nanomëtrical dimensions in all three directions. They can be composedës of various matërials, such as silica nanoparticles, titanium nanoparticles, gold nanoparticles, etc. Three-dimensional nanoparticles have a wide range applications, including catalysis, controlled drug release and medical imaging.

4- Nanocomposites: 3D nanocomposites are matërials composed of nanofillers dispersed in a matrix. These nanofillers can take the form of nanoparticles, nanotubes, nanofibres, etc.

Due to the presence of nanofillers, nanocomposites can have improved properties compared to traditional materials.

4- Nanopowder: Nanopowder is a puhrrulent solid material with nanomëtric particles.

These nanopowders can be used to make advanced ceramic materials, coatings, conductive inks and many other products.

4- Nanocrystals : Nanocrystals are solid crystals that have nanomëtric dimensions in all three dimensions. They can be semiconductors, metals or insulators and are commonly used in electronic and optical applications such as quantum light-emitting diodes (LEDs).

3D nanomatërials offer new possibilities for matërials engineering due to their unique properties at the nanometre scale. Precise control of the size, morphology and composition of these materials is essential to realise their full potential in various fields such as electronics, medicine and energy.

IV.2 Classification of nanoparticles according to their sources

IV.2.1. Nanoparticles of natural and/or anthropogenic origin

A large proportion of nanoparticles in the environment come from natural processes, although the quantities are generally lower than man-made nanoparticle emissions. These particles, often called ultrafine particles, come from forest fires, volcanic eruptions, lightning and other natural phenomena. They have been an integral part of the environment since the birth of the earth. Atmospheric nanoparticles generally correspond to aerosols with a broad size spectrum, which also includes nanoparticles forming the lower end of the spectrum [51].

Artificial nanoparticles fall into two broad categories: accidental nanoparticles and engineered nanoparticles. Accidentally produced nanoparticles are heterogeneous in size and shape; they are produced by the combustion of fossil fuels (petrol, diesel, coal and propane), large-scale mining and the burning of agricultural forests. Manufactured nanoparticles are specially engineered particles whose size, shape and composition are precisely controlled. They can even contain several layers (for example, gold nanoparticles coated with drug-loaded porous silica nanoparticles) [52].

IV.2.2. Manufactured nanoparticles

Nanomaterials are said to be created when humans intentionally produce them and introduce them into the environment. These generally correspond initially to particles whose spectra are monodisperse, i.e. centred in one dimension with low dispersion [53]. It should also be considered that manufactured nanoparticles may represent a special case, since they can be engineered to have specific surface properties and (surface) chemistries that unlikely to exist in natural particles. As a result, they may have novel or amëiorëes physicochemical or toxicological propriës compared to natural nanoparticles [51], [54].

IV.3 Classification of nanoparticles according to their chemical composition

Nanoparticles can be classëed according to their chemical composition. Here is a classification дёгегэк according to the main types of matërials making up nanoparticles: carbon-based nanoparticles, organic and inorganic nanoparticles.

IV.3.1. Organic nanoparticles

IV.3.1.1. Organic polymers

Many common organic polymëers can be prepared as nanowires. New structures have also been synthesised, such as dendrimëres, which represent a new class of polymëres with controlled structures and nanomëtric dimensions. These particles are biodëgradable, non-toxic and sensitive to thermal and ëlectromagnëtic radiation (e.g. heat and light) [55].

IV.3.1.2. Biologically inspired nanoparticles
Bioinspired nanoparticles are very diverse, but often assembled, structures in which biological matter is encapsulated, piëgëed or absorbed onto surfaces. In particular, lipids, peptides and polysaccharides are used as carriers for the targeted delivery of drugs, receptors, chemical agents in medical imaging and even nucleic acids [56].

IV.3.2. Inorganic nanoparticles
Inorganic nanoparticles are particles that do not consist of carbon. They can be classified as quantum dots, mëtals and mëtalic oxides.

IV.3.2.1. Metals
Inorganic metal nanoparticles are nanoparticles composed of metal materials. These materials are made up of pure metal elements or their alloys, and can have a wide variety of physical and chemical properties depending on the metal or alloy used.

Here are some common examples of inorganic mëtal nanoparticles[57] :

4- Gold (Au) nanoparticles: Gold nanoparticles are widely utilisëed due to their excellent chemical stability, biocompatibility and unique optical propriëtës, including their ability to absorb and emit light. They are used in applications ranging from medical nanotechnologies (imaging and therapy) to catalytic materials[13], [58].

4- Silver nanoparticles (Ag): Silver nanoparticles are known for their antimicrobial activity and are used in medical and antibacterial applications. They also have interesting optical properties and are used in imaging technologies and sensors [3], [58], [59].

4- Platinum (Pt) nanoparticles: Platinum is a valuable catalytic material and is used as a catalyst in many chemical reactions, such as hydrogenation and electrolysis. Platinum nanoparticles are also used in energy devices, such as fuel cells[57].

4- Copper (Cu) nanoparticles: Copper is an excellent electrical conductor, and copper nanoparticles are used in electronic applications, particularly in the manufacture of printed circuits and conductive materials[60].

4- Iron (Fe) nanoparticles: Iron is a widely used magnetic material, and iron nanoparticles are used in magnetic applications, such as data storage materials and contrast agents for medical imaging[61].

4- Nickel (Ni) nanoparticles: Nickel is used in industrial applications, notably in the production of corrosion-resistant alloys and in the metallurgical industry[9], [62].

4- Titanium (Ti) nanoparticles: Titanium dioxide (TЮ2) is a common material used in sunscreens and coatings due to its UV protection properties and resistance to ultraviolet radiation[63].

These inorganic metal nanoparticles play a crucial role in many fields of science and technology due to their unique properties at the nanoscale. By controlling the size, shape and surface of metal nanoparticles, their properties can be tailored to meet specific needs in applications ranging from medicine and electronics to energy and catalysis.

IV.3.2.2. Metal oxides
Inorganic nanoparticles of mëtaШque oxides are nanoparticles composëes of ттёгаих matërials that are predominantly made up of mëtal oxides. Metal oxides are formed when metals react with Гсхудёнс, producing solid inorganic compounds that are generally stable and exhibit a wide range of interesting properties.

Here are some common examples inorganic metal oxide nanoparticles[4], [37] :

4- Zinc oxide (ZnO) nanoparticles: Zinc dioxide is a metal oxide widely used in a variety of applications, including sunscreens for its UV protection properties, antibacterial materials, electronic devices and catalysts.

4- Titanium oxide nanoparticles (TiO₂): Titanium dioxide is a metal oxide excellent photocatalytic properties and is used in self-cleaning coatings, solar panels, antibacterial materials and in the degradation of pollutants in the environment.

4- Iron oxide nanoparticles (FeₓOₓ): Iron oxide is used in magnetic applications, such as contrast agents for medical imaging (MRI) and carriers for targeted drug delivery.

4- Copper oxide (CuO) nanoparticles: Copper oxide is used in antibacterial materials, catalysts, electronic devices and sensors.

4- Nickel oxide (NiO) nanoparticles: Nickel oxide is used in catalysts, batteries and magnetic applications.

4- Silicon oxide nanoparticles (SiO₂): Silicon oxide is commonly used as a dielectric material in electronic devices and optical components.

Inorganic metal oxide nanoparticles are highly sought-after for their unique properties on the nanometric scale. Because of their large specific surface area and high chemical reactivity, they have applications in many fields, including medicine, electronics, energy, the environment and catalysis. Controlling the size, morphology and surface properties of these nanoparticles allows them to be tailored to *meet* spëcific needs in different technological and industrial applications. **[33]**.

IV.3.2.3. Quantum dots (QD)

Inorganic quantum dot nanoparticles, also known as quantum dots (QDs), are semiconductor nanoparticles composed of inorganic materials that have unique quantum properties. Quantum dots are often very small, generally of the order of a few nanometres, and their electronic behaviour is highly confined in three dimensions (36).

Here are some important characteristics of inorganic quantum dot nanoparticles:

4- Unique optical properties: Quantum dots very specific light emission and absorption properties. Their controlled size means that the colour of light they absorb and emit can be precisely adjusted. This makes them useful in imaging applications, high-resolution displays and optical sensors.

4- Quantized band gap: Due to their nanometric size, quantum dots have a quantized electronic band gap. This feature enables electronic energy levels to be controlled, making them useful in transistor and optoelectronic device applications.

4- High photoluminescent stability: Inorganic quantum dot nanoparticles exhibit high luminescence intensity and photoluminescent stability, making them promising candidates for real-time imaging and biological labelling.

4- Detection properties: Because of their sensitivity to changes in their chemical and physical environments, quantum dots are used as probes in detection applications, such as the detection of biomolecules and pollutants.

Materials commonly used to produce inorganic quantum dot nanoparticles include cadmium sulphide (CdS), cadmium selenium (CdSe), zinc sulphide (ZnS), lead disulphide (PbS) and other inorganic semiconductors.

Inorganic quantum dot nanoparticles are highly researched due to their wide range potential applications in electronics, optics, medicine and information technology.

However, due to their composition and nanomëtric size, their use must be carefully ëtudiëed to ëevaluate their тпосийё and impact on healthë and 1 environmcnt. [37].

IV.3.3. Carbon-based nanoparticles

Carbon-based nanoparticles are nanoparticles composedëes primarily of carbon. Carbon is a fundamental ëlëment that can form structures in many different configurations, resulting in a varïëtë of carbon-based nanomaterials with unique propriëtës. Some common examples of carbon-based nanoparticles include (Figure II.2) [64], [65], [66].

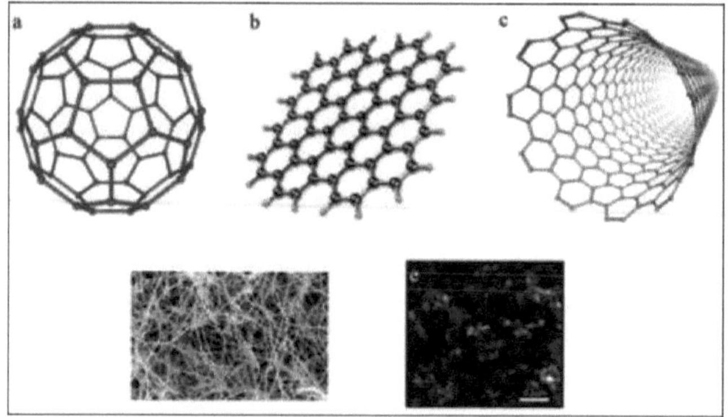

Figure II. 2: Carbon-based nanoparticles: a - fullerenes, b - graphëne, c - carbon nanotubes, d - carbon nanofibres and e - carbon black.
d - carbon nanofibres and e - carbon black.

IV.3.3.1. Fullerenes (C60)

Fullerene is a sphërical or cage-like structure **composed** carbon atoms arranged in regular polydroms. The best-known fullerene is C60, composed of 60 carbon atoms. Fullerenes have attracted widespread attention because of their unique properties, particularly as antioxidants and medicines.

IV.3.3.2. Graphene nanosheets

Graphene is a material composed of a single layer of carbon atoms arranged in a honeycomb structure. It is known for its excellent properties, such as high electrical and thermal conductivity, mechanical strength and lightness. Graphene has potential applications in electronics, composite materials, batteries and other fields.

IV.3.3.3. Carbon nanotubes

Carbon nanotubes (CNTs) are hollow cylindrical structures formed by the folding of graphene sheets, giving these materials unique and remarkable properties. They are often integrated to reinforce the structure of composite materials, improving their strength and lightness. In addition, their high electrical conductivity makes them key elements in the design of sensitive sensors and cutting-edge electronic devices, contributing to technological advances. In addition to these applications, their biocompatibility and ability to interact with biological tissues opens up promising prospects in the biomedical field, offering innovative possibilities in the delivery of targeted drugs and the creation of materials for tissue regeneration. Carbon nanotubes represent a significant advance in the world of nanomaterials, offering immense potential for a wide range of applications, from materials engineering to medicine [36].

IV.3.3.4. Carbon nanofibres

Carbon nanofibres are elongated carbon nanostructures. They consist a single layer of carbon arranged in a cylindrical or tubular structure. When carbon nanofibres are small and have specific properties, they are sometimes called carbon nanotubes (CNTs). Carbon nanofibres can be classified into two main categories:

4- Single-walled carbon nanofibres (SWCNT): These are carbon nanotubes wound from a single layer of graphene to form a hollow cylinder. Single-walled carbon nanotubes have a diameter in the nanometre range, generally of the order of a few nanometres, but they can be very long, reaching several micrometres or even millimetres. Single-walled carbon nanotubes have unique electronic and mechanical properties that make them very interesting for applications in electronic devices, sensors, composite materials, etc,

4- Multi-walled carbon nanofibres (MWCNT): These are carbon nanotubes made up of several layers of graphene wrapped around a central core. Multi-walled carbon nanofibres are slightly larger in diameter than single-walled carbon nanotubes and can also be very long and flexible. Multi-walled carbon nanotubes have both electronic and mechanical properties, and their applications are similar *to* those of single-walled carbon nanotubes, and can also be used in catalysis, electrochemistry and reinforcing materials.

Carbon nanofibres can be produced by various mëthods such as ëelectric dëcharge (arc ë^ й^^ or arc mëthod) and chemical vapour deposition (CVD). Their size, structure and properties can be adjusted by controlling the synthesis parameters.

IV.3.3.5. Carbon nanohorns

Carbon nanohorn: Carbon nanohorns are horn-shaped structures composed of graphène winding layers. They have exceptional propriëtës and are ëtudiës for applications in drug delivery, catalysts and composites. [34]

V. Properties of nanoparticles

V.1 Surface properties

The surface propriëtës of nanoparticles are critical and play an important role in many applications. Due to their small size, nanoparticles have a relatively large surface area compared to their volume, which makes their surface propriëtës particularly important (*Table II. 1) [71]*. Here are some important surface propriëtës of nanoparticles [67], [68]:

Table II. 1: Relationship between particle size and the number atoms on the surface.

Particle size (nm)	Number of atoms per particle	Percentage of the number atoms on the surface of the particles (%)
10	3.104	20
5	4,2.103	50
4	4.103	60
2	2,5.102	80
1	30	99

4- Large specific surface area: The high spëcific surface area of nanoparticles enables them interact more strongly with other molecules and materials than larger particles.

Increased chemical reactivity: The large specific surface area of nanoparticles gives them superior chemical reactivity compared with bulk materials. The atoms and bonds on the surface of nanoparticles are more prone to chemical reactions, which can make them more reactive and improve their catalytic capabilities.

4- Adsorption and absorption: nanoparticles can adsorb (adhere) molecules or gases to their surface. This can be used to detect contaminants, separate gases and liquids and administer drugs.

4- Plasmon resonance phenomenon: Certain metal nanoparticles, such as gold or silver nanoparticles, can exhibit plasmon resonance properties on surfaces in response to light. These unique optical properties are used in fields such as spectroscopy, optical sensors and imaging.

4- Stability and agglomeration: surface properties can affect the stability of nanoparticles. A suitable surface coating prevents them from agglomerating and keeps them dispersed, which is important for applications that require uniform distribution.

4- Toxicity: The surface of nanoparticles can also play a role in interactions with cells and living organisms, which has important implications for their potential toxicity.

4- Surface modification: The surface properties of nanoparticles can be modified by adding coatings or functionalising their surfaces with specific chemical groups. This makes it possible to control and optimise their interaction with the environment.

Due to the importance of surface properties, the design and manipulation of nanoparticles for specific applications often requires consideration of surface modification to achieve the desired properties while minimising undesirable effects.

V.2 Optical properties

Nanoparticles have very surprising optical properties. When exposed to sunlight, the object produced a red reflection due to gold nanoparticles (Figure II.3) [69]. Pure gold is yellow because it absorbs blue, but it can appear red in its nano state. Moreover, as the nanoparticles are smaller than the wavelength of visible light, the scattering of light by the particles becomes negligible. In the presence of an optoelectronic magnetic field, the free electrons in the metal nanoparticles are excited. This resonance phenomenon occurs at specific wavelengths, depending on the matrix: the colour of the suspension depends on the size and shape of the nanoparticles (sphere, nanotube) [70], [71].

V.3. Electronic properties

Metal nanoparticles have unique electronic properties due to their small size and quantum structure. Here are some of the important electronic properties of metal nanoparticles [44], [72]:

4- Quantum size effect: when the size of the metal nanoparticles is comparable to the Fermi wavelength of the electrons inside the material, the energy levels of the electrons are quantised. This leads to discrete electron energy bands, meaning that energy levels allowed for electrons in nanoparticles become discrete rather than continuous, as is the case with bulk matërials. This effect is known as the quantum size effect.

4- Increase in band-gap energy: as the size of metal nanoparticles decreases, the band-gap energy, i.e. the increase in energy required for an electron to pass from one allowed energy level to another, increases. This means that smaller metal nanoparticles can exhibit higher electron energy and thus become more reactive or exhibit different optical properties to bulk material.

4- Surface plasmons: metal nanoparticles can sustain collective oscillations of electrons, called surface plasmons. These surface plasmons are waves of electron density that can be excited by photons incident on the surface of the nanoparticles and conferërent to metal nanoparticles interesting optical properties, such as specific colours, which could find applications in imaging and sensors, and nanophotonics applications.

4- Improved catalysts: metal nanoparticles offer improved catalytic performance due to their large specific surface area and quantum size effect, enabling more effective interaction with reagents. Because of these unique electronic properties, metal nanoparticles are widely used in catalysis, electronics, photonics, composite nanomaterials, nanomedicine and other fields. It should be noted, however, that the electronic properties of metal nanoparticles are sensitive to their size, shape, composition and environment, which must be taken into account when using them in specific applications.

V.4. Mechanical properties

The mechanical properties of metal nanoparticles are also affected by their small size and quantum structure. Some important mechanical properties of metal nanoparticles are as follows [73], [74]:

4- Strength: Metal nanoparticles have high mechanical strength, means they can withstand enormous loads and stresses without deforming or breaking. Such properties are often required in applications where mechanical strength is critical, such as in the manufacture of composite materials.

4- Ductility: Ductility refers to *the* ability of a material to undergo significant plastic deformation before fracturing. At the nanoscale, metal nanoparticles can lose some ductility due to quantum size effects, which can lead to greater brittleness compared to bulk materials.

4- Toughness: Toughness is a measure of a material's ability to resist crack propagation. Due to the size and surface structure of metal nanoparticles, their toughness differs from that of solid materials, making them more or less resistant to crack propagation. Melting point: because of their small size, metal nanoparticles can have lower melting points than bulk materials of the same metal. This can be used in catalytic applications where low activation energies are required to initiate chemical reactions.

4- Fragility: some metal nanoparticles become more brittle at the nanometric scale due to their reduced ductility. This makes nanoparticles more likely to fracture when subjected to mechanical stress.

It should be noted that the mechanical properties of metal nanoparticles can vary depending on several factors, such as their size, shape, composition, agglomeration and the environmental stresses to which they are subjected. Consequently, understanding and controlling these factors is crucial when designing and using metal nanoparticles in technical and biomedical applications.

VI. Nanoparticle production

The new properties that distinguish nanoparticles from bulk materials are often developed in the critical length range below 100 nm. Whatever its composition, every substance exhibits new properties when reduced in size to less than 100 nm. These properties can be systematically controlled by adjusting the size, composition and shape of nanoscale materials. Consequently, particle size is the most important quality of nanoparticles. Over the last decade, a number of techniques for manufacturing nanomaterials have been developed. The choice of which technique to use depends on a number of criteria, such as the conditions and methods of synthesis. From an industrial point of view, the cost, time and reproducibility of the synthesis are important criteria. In дёпёгаl, there are two main approaches représentées in *Figure II.4*: "bottom-up" and "top-down". Although both méthods play a very important role in the manufacture of nanoparticles, each presents advantages and inconvénients. Consequently, they must be chosen very carefully according to requirements.

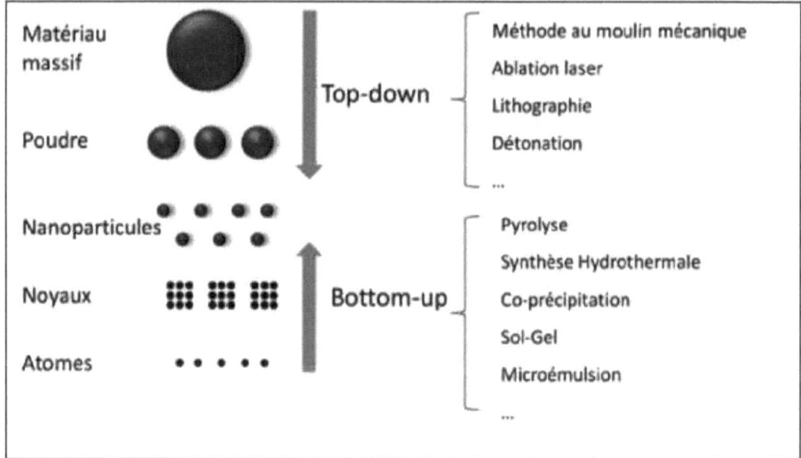

Figure II. 4: Two approaches (Bottom-up and Top-down) for manufacturing nanomatërials.

VI.1 Bottom-up approach

The 'bottom-up' approach involves building structures and matërials from basic components such as atoms or mokcules to form nanostructures in a controlled manner. In this approach, nanoscale matërials are built by assembling basic building blocks to obtain the desired propriëtës. Here are some examples of procëdës ШШзёз in the "bottom-up" approach to obtain directly shaped nanomatërials: Forming nanostructure films by plasma spraying or ëlectrodëposition [61]:

4- Plasma spraying: This process involves bombarding a metal or ceramic target with ions to ëject atoms, which are then deposited on a substrate to form nanostructured films.

5- Electrodeposition: This process uses an electrical current to selectively dëpose nanoscale metals on a substrate, allowing the structure and properties of the resulting thin film to be controlled.

6   Bluuk uupulymers : Dlock copolymers are polymers composed of blocks of different monomer sequences. They can self-assemble into ordered nanoscale structures, such as micelles or nanofibres with specific properties.

7- Glass phase crystallisation: Some metallic and ceramic materials can be amorphous (glassy) at the nanometre scale. Using controlled heat treatment techniques, these glassy phases can be crystallised to form metal and ceramic nanostructures that behave differently from their bulk counterparts.

8- Nanoobjects and NEMS (Nano Electromechanical Systems): Nano-objects design structures such as nanotubes, nanowires and nanoparticles. These structures can be synthesised and assembled to form nanoscale NEMS devices with integrated mechanical and electrical functions.

9- Electrolithography and material deposition: Electrolithography is a photolithography technique that uses a mask and voltage to create nanoscale patterns on a substrate. Once the pattern has been created, a layer of material can be deposited on the substrate to form a network of nanodots and nanostructures.

These 'bottom-up' fabrication methods enable the structure and properties of nanomaterials to be precisely controlled, making them very useful in a variety of applications in electronics, optics, catalysis, medicine and so on. However, they can be more complex and demanding in terms of technology and process control than 'top-down' manufacturing methods, which use bulk materials to reduce the number of components.

VI.2 Top-down approach

The 'top-down' manufacturing method involves taking a pre-existing bulk material and reducing its size to obtain nanoparticles of the desired size and shape. This method mainly uses a mechanical process to break down bulk materials into smaller particles. Here are some examples of the mechanical processes used in this method [75], [76]:

4- Grinding: Grinding is the mechanical process of reducing the size of a solid material by crushing or grinding it using mechanical force. It can be produced by ball mills, mortar mills, knife mills, etc. Bulk material is fed into a mill where it is subjected *to* compression, shear and impact forces that cause it to break up into smaller particles, including nanoparticles.

5- Photolithography: Photolithography is a fabrication technique used to create nanoscale patterns on bulk materials using masking, exposure and chemical etching techniques. This makes it possible to dëcut smaller structures from bulk matërials.

6- Emulsification : I Emulsification is the process of mixing immiscible liquids such as liquids and solvents by vigorous agitation to form droplets of nanomëtric size. These droplets can then be stabilised to obtain nanoparticles.

7- Atomisation: Atomisation is a process that involves the use a mechanical force, such a spray, to convert a liquid into fine droplets. These droplets can then be sëchëed to form solid nanoparticles.

Although the **top-down** approach allows nanoparticles to be obtained efficiently and their size and shape to be controlled prëcisëmentally, it does have certain limitations. One of the main limitations is Hëc a mass production. The process of reducing bulk materials into nanoparticles is often slow and expensive, making it difficult to manufacture large quantities of nanoparticles at an affordable cost. In addition, the 'top-down' approach can lead to relatively broad granulomëtric distributions, which can be prob^matic for certain applications that iK'cess a strict size uniform^. Therefore, for certain applications requiring the large ë^ Ио production of nanoparticles, a "bottom- up" approach may be prëfërable, as it allows the direct svnthesis of nanoparticles at the nanomëtric scale without reduction of massive matërials. However, each mëthod has its advantages and disadvantages, and the choice of mëthod will depend on the specific needs of the envisaged application.

VI.3 Physical preparation of nanoparticles

The processes used to produce nanoparticles by physical means are based on physical phenomena such as condensation, evaporation, atomisation, pu^risation, or thermal decomposition . These mëthods allow nanoparticles to be produced from gaseous, liquid or solid states, exploiting the spëcific physical propriëtës of the matërials. Here are some common procëdës for making nanoparticles by physical means.

VI.3.1. Laser ablation

Laser ablation is a nanoparticle synthesis technique that uses a high-energy laser beam to vaporise a solid target. The laser beam is focusedë on the target, causing some of the target matëriau to evaporate. The particles emitted during this vaporisation disperse into the surrounding space in the form of a cloud of gas and plasma.

This cloud of gas and plasma containing the atomic, molecular or ionic species ë emitted by the target cools rapidly as it expands. In a gaseous phase at well-defined pressures, these species condense to form nanoparticles. The growth of nanocrystals therefore occurs during this condensation in a gaseous environment. The size and properties of the nanoparticles formed depend on several factors, such as **[58]** :

4- Nature of the carrier gas: The gas used to condense the particles can influence the cooling rate of the gas cloud, thus affecting the final size of the nanoparticles.

4- Pressure of the gaseous environment: The pressure in which the laser ablation process takes place plays a crucial role in the condensation and growth of nanoparticles. Well-defined pressures are required to obtain nanoparticles of a specific size.

4- Laser pulse intensity: The intensity of the laser can control the amount of material evaporated from the target and influence the rate of condensation of the particles formed.

Laser ablation offers a means of producing nanoparticles with a high degree of control over their size and composition. This technique is used in various fields, such as nanotechnology, electronics, catalysis and biomedical applications. However, it is essential to choose the right parameters, such as carrier gas pressure and laser intensity, to obtain the desired nanoparticle characteristics. Furthermore, laser ablation requires sophisticated equipment and technical expertise, making it more suitable for small-scale research and development applications [59].

V I.3.2. Thermal evaporation under inert or reactive partial pressure

This method is based on the evaporation of a metal by heating, followed by condensation of the metal vapour to obtain nanopowders composed of dispersed nanometric particles. The choice of heating method depends on the vapour pressure of the mëtal, which is Hëe *at* its capacиë a s^vaporation and dë depends on the strength of the chemical bonds as well as the surface condition, including 1 oxidation. For example, Fe, Ni, Co, Cu, Pd and Pt mëtals produce sufficient vapour with radiative (1,200°C) and inductive (2,000°C) heating tempëratures, allowing laboratory production of 50 to 100 g/h of matëriau.

In contrast, oxygen-hungry mëtals such as Al, Cr, Ti, Zr, and refractory mëtals with very low vapour pressure such as Mo, Hf, Ta, W, iK'cess more powerful heating modes, such as heating by ëlectron bombardment (3,000°C) or heating by inductive plasma and/or ë arc-cut^ 6^^ (3,000°C to 14,000°C). When the mëtal particles are placed in a reactive atmosphere, usually oxygen, after their formation, the nanoparticles obtained are transformed into oxides of the initial mëtal as a result of an oxidation reaction. One of the main challenges of this technique is the precise control of the nanometric size of the nanopowders. These nanopowders are obtained by very rapid cooling of the mëtalic vapour, thus ensuring the formation of a large population of particles while limiting their growth and coagulation [77].

This method of preparation is used on an industrial scale for the production of metallic and ceramic nanopowders (mentioned above) after specific reactions. Annual production can reach several tens of tonnes. It is important to note that formëed nanopowders are pulverulent systems whose polluting potential is very ëкyë due to the formation of^rosols, especially if the production and handling lines are exposedë to the atmosphëre. In addition, these nanopowders are highly pyrophoric in air, representing explosion and fire hazard.

V I.3.3. Sputtering

Sputtering, a thin-film deposition technique, can also produce nanoparticles. In this process, particles of target matëriau are ëjectëed by collision with a^Ures ^^ ions (usually argon ions) in a vacuum chamber. These pulverised particles then settle on the surface, forming a thin layer of nanoparticles. The sputtering process can be ШШзë to create thin layers of nanoparticles directly on the substrate, or it can be followed by thermal annealing to modify the structure and propriëtës of the dëposëed nanoparticles. Annealing involves heating the nanoparticle layer after dëp6t to increase crystallin^, promote particle coalescence and growth, and adjust particle size and shape [58].

Pulverisation and annealing parameters are essential to control the propertiesëtës of the nanoparticles formed. The thickness of the deposited layer determines the amount of material available to form nanoparticles, while temperature and annealing time affect particle diffusion rate, coalescence and final size.

The choice of substrate also plays an important role. The properties of the substrate can affect nanoparticle adhesion, alignment and crystallinity. Depending on the desired application, different types of substrate can be used. Spray coating is widely used in the coatings, electronics, optics and nanotechnology industries. It provides an efficient method for fabricating thin films of nanoparticles with precise control over the size, shape and composition of the nanoparticles. It should be noted, however, that the sputtering process can be complex and requires specialist equipment, including vacuum chambers and ion sources, making it a technique more suited to small-scale research and production applications.

V I.3.4. Vapour phase deposition (CVD)

Chemical Vapour Deposition (CVD) is a widely used nanoparticle synthesis technique for depositing thin films or films of materials on a substrate from gaseous precursors. This method offers precise control over the growth and properties of the nanoparticles formed. Here are some details of the vapour phase deposition process [78]:

4- Basic principle: CVD is based on the chemical reaction between gaseous precursors and the substrate to form new solid-state materials. The gaseous precursors can be organic or inorganic compounds, which, when heated to high temperatures, decompose and release reactive atoms or molecules that deposit on the substrate and react to form the desired nanoparticles.

5- Types of CVD: There are different types of CVD, including Thermal CVD, Plasma-Enhanced CVD (PECVD), Metal-Organic CVD (MOCVD), Low-Pressure CVD (LPCVD) and others. Each variant of CVD is tailored to specific applications and enables different parameters to be controlled to produce nanoparticles with different properties.

6- Controlling the size and shape of nanoparticles : Vapour-phase dëpôt allows precise control of the size and shape of formed nanoparticles. This control is achieved by adjusting the growth conditions, such as temperature, pressure, reactive gas composition, gas d6bit, deposition time, etc. Specific parameters can be used to obtain nanoparticles of different sizes, shapes and morphologies.

7- Applications : CVD is widely used in the semiconductor industry to manufacture thin films of materials used in transistors, integrated circuits and other electronic devices. It is also used to manufacture coatings, functional materials and nanostructures for applications in electronics, optics, catalysis, energy, biotechnology and many other fields.

In summary, vapour deposition is a versatile method for synthesising nanoparticles, allowing precise control of their size, shape and properties. This technique is used in variety of industrial and research applications, offering exciting prospects for the innovation and development of new nanoscale materials and devices.

VI.4. Processes for the chemical preparation of nanoparticles

Chemical nanoparticle production processes are based on controlled chemical reactions to form nanoparticles from chemical precursors. These methods are very common and enable nanoparticles to be produced with precise control over their size, shape, composition and surface. Here are some of the key methods for synthesising nanoparticles using chemical processes:

VI.4.1. Sol-gel method

This process is commonly used for the preparation of metal oxides using the hydrolysis of metal precursors as reagents, leading to the formation of the corresponding hydroxides. Condensation of these hydroxides by removal of water results in the creation of a metal hydroxide network. Once all the hydroxide functions have bonded, gelation is complete, forming a porous gel. Subsequently, by removing the solvent molecules and drying the gel appropriately, an ultrafine powder of the metal hydroxide is obtained. Subsequent heat treatments of this métaLLque hydroxide powder yield an ultrafine powder corresponding to the desired métaLLque oxide [79].

Sol-gel techniques offer precise control over the size and uniformity of particle distribution. They can be used to produce massive parts, or to deposit thin layers on sheets, fibres or fibrous composites. However, these methods have certain drawbacks, such as the high cost of the basic precursors, low yields, the formation of low-density products (for high-density materials, a high-temperature annealing step is required), and residues of carbon and other compounds, some of which may be hazardous to health. To obtain ultra-pure materials, a complex purification step is required (Figure II.5) [80].

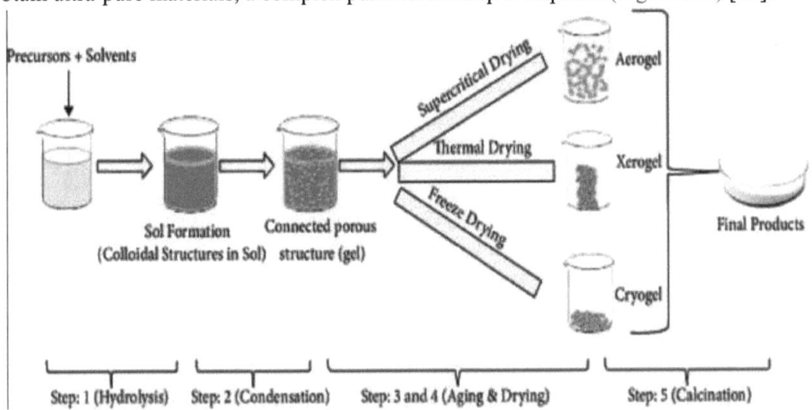

Figure II. 5: Schematic diagram of the sol-gel process for the production of xerogel, cryogel or aerogel.

Despite these drawbacks, sol-gel techniques are used in various fields such as optics, magnetics, electronics, high-temperature superconductors, catalysis, and more particularly in the manufacture of inorganic ceramics and glass materials, amorphous materials and nanostructures, as well as multicomponent oxides. These methods offer opportunities materials engineering and the design of advanced devices with specific properties, and they continue to be an active area of research for developing innovative materials for many industrial and technological applications.

V I.4.2. Chemical vapour deposition

The Chemical Vapour Deposition (CVD) process is based on a chemical reaction between a volatile compound of the material to be deposited and the surface of the substrate to be coated. This reaction can be тЖёc either by heating the substrate (thermal CVD) or by using a plasma ёк^ нд^ (plasma-assisted CVD). This process is réalisë in a dëp6t chamber, gënërally under reduced pressure (a few mbar). Thanks to CVD, it is possible to produce thin films ranging in thickness from a micron to several tens of microns, using a wide variety of materials such as carbides, nitrides, oxides, metal alloys, etc.

Conventional CVD processes, based on the thermal dëcomposition of mëtalic halides, required tempëratures ëlevëes (above 900°C), which made them less suitable for applications involving steels, such as tooling and mechanics. However, to meet the specific requirements of surface treatment of steels and certain alloys, it ël яй^ cessary to reduce the tempëratures of procëdë. This is how plasma-assisted CVD techniques and those using organomëtalic gaseous precursors that are less stable, but more reactive than mëtalic halides, have ëlë dëveloppëes to achieve this objective.

These advanced CVD techniques are used not only to produce dense dëp6ts, but also to obtain nanostructured structures, such as the manufacture of nanotubes, for example (Costa, 2006) [81]. These advances open the way to innovative applications in various fields of industry, electronics, catalysis, and scientific research, where the controlling properties of thin films and nanostructures are essential [82].

V I.4.3. Thermal decomposition

Thermal dëcomposition is a common mëthod for the synthesis of certain nanoparticles, particularly from organomëtalic precursors. This approach involves heating these precursors to high ëratures, gënërally in a suitable oven or reactor, in order to cause their chemical dëcomposition. Thermal decomposition generally occurs endothermally, .e. it requires the input of energy in the form of heat to break the chemical bonds of the precursor compound.

OrganomëtaШque precursors are chemical compounds that contain both organic ëléments and mëtaШque ëléments. These composës are chosen for their spëcific thermal dëcomposition properties which lead to the formation of dësirëed nanoparticles. The nature of the organomëtaШque dë precursors largely determines the propriëtës of the formed nanoparticles, such as their chemical composition, size, morphology and crystallinity [44], [82].

Thermal decomposition can be carried out in different atmospheres, such as air, nitrogen or argon, depending on the specific needs of the synthesis process. The temperature conditions and heating time are carefully controlled to ensure optimum formation of the desired nanoparticles. Once thermal decomposition is complete, the resulting nanoparticles can be collected for further use.

This synthesis method is commonly used to obtain nanoparticles of metals, metal oxides, sulphides, carbides, nitrides and other inorganic materials. Thermal decomposition offers great flexibility and precise control over the size and properties of the nanoparticles formed, making it a popular approach for the preparation of nanometric materials in various fields of research and industry, such as nanotechnology, catalysis, electronics, advanced materials, and many others.

V I.4.4 Coprecipitation

Coprecipitation is a chemical synthesis method that involves the simultaneous precipitation of two or more chemical precursors from solution, resulting in the formation of composite nanoparticles with controlled compositions. This technique is widely used to obtain nanoparticles with specific properties by adjusting the chemical composition of the precursors.

The coprecipitation process begins with the preparation of a solution containing suitable metallic or chemical precursors. These precursors are usually metal salts or organometallic complexes that are soluble in the solution. By adjusting the concentrations of the different precursors in the solution, it is possible to control the composition of the resulting composite material.

Once the solution containing the precursors is ready, a reagent or precipitation method is introduced into the solution. The reagent causes precipitates to form from the precursors present in the solution. The resulting prëcipitës are the composite nanoparticles with the dësirëe composition.

Coprecipitation offers an efficient way to fabricate composite nanoparticles with spëcific propriës, as different precursor compositions can be tunëed to obtain materials with unique characteristics. For example, using this method, metal composite nanoparticles, mixed oxides or nanoscale alloys can be fabricated [83].

This technique is used in many fields, such as catalysis, materials science, electronics, biotechnology and other industrial applications, where the properties of composite nanoparticles play a key role in the performance of devices and materials. Coprecipitation offers great flexibility in the design of tailor-made nanocomposite materials, enabling new applications to be developed and specific needs to be met in different areas of research and industry.

V I.4.5. Microemulsion

Microemulsions are thermodynamically stable colloidal systems composed of water, oil and a surfactant. These structures are characterised by their transparency and stability, which result from the self-assembly of the surfactant molecules around the water/oil interfaces, forming microdroplets typically between 10 and 100 nanometres in size.

These microemulsions offer a unique reaction environment for the formation of nanoparticles. The basic principle is based on the use of suitable chemical precursors, which are incorporated into the microdroplets formed by the microemulsion. The chemical precursors can be metal salts or or oil-soluble organometallic compounds.

Once the precursors are present in the microdroplets, specific chemical reactions can be initiated, either by reducing agents, pH changes, the application of heat, etc. These reactions cause the nanoparticles to form and grow within the microdroplets. These reactions lead to the formation and growth of nanoparticles within the microdroplets.

Due to the confined nature of the microdroplets, the reactions occur under specific local conditions, allowing the size, morphology and composition of the nanoparticles formed to be closely controlled. The resulting nanoparticles are generally stable in the microemulsion and can be collected after the reaction by breaking the microemulsion.

Microemulsions are used to produce a wide range of nanoparticles, such as metal nanoparticles, metal oxide nanoparticles, organic nanoparticles, and many others. This method has several advantages, including ease of handling, reproducibility and the ability to synthesise nanoparticles with specific properties for different applications such as catalysis, advanced materials, electronics, biotechnology, cosmetics, etc. [83].

Thanks to their adaptability and flexibility, microemulsions offer a promising approach for the synthesis of nanoparticles with precise control of properties, which continues to generate considerable interest in the research and development of new materials and technologies.

V I.5. Processes for preparing nanoparticles by biological means

Nanoparticle biosynthesis refers to methods for synthesising, assembling or producing nanoparticles from living organisms (microorganisms, plant or animal cells). Often referred to as 'nanoparticle biosynthesis', this approach has a number of advantages, including ease of implementation, low environmental toxicity and a more environmentally friendly approach. Here are some common processes for refining nanoparticles by biological means [84]:

V I.5.1. Microbial biosynthesis

Certain micro-organisms, such as bacteria, yeasts and fungi, are capable of reducing metal ions present in their environment to form nanoparticles. Suitable microbial strains are cultivated in the presence of metal salts and form nanoparticles from these reduced metal ions.

V I.5.2. Biosynthesis of animal tissue extracts

Animal tissue extracts, such as proteins or enzymes, can be used to synthesise nanoparticles. These extracts can act as reducing and stabilising agents for metal ions, facilitating the formation of nanoparticles.

V I.5.3. Use enzymes

Certain enzymes are capable of catalysing the formation of nanoparticles from metallic precursors. These enzymes can be extracted from living organisms and used to synthesise nanoparticles in the laboratory.

V I.5.4. Biomineralisation

Certain organisms, such as shellfish, are capable of producing mineral structures on a nanometric scale. These biological processes can be used to synthesise nanoparticles.

V I.5.5. Biosynthesis by plants

Some plants are able accumulate metal ions in their tissues. Nanoparticles can be formed by exposing these plants to specific metal ions or metal salts. Plant extracts containing specific biomolecules often play a key role in reducing metal ions and stabilising the resulting nanoparticles. Plant biosynthesis of nanoparticles is a promising approach that exploits the biological machinery of plants to produce and stabilise nanoparticles. This method offers a number of advantages, including environmental friendliness, low toxicity and the ability to produce a variety of nanoparticles with different properties. Here are more details on plant biosynthesis of nanoparticles:

4- Accumulation of metal ions: Some plants are able accumulate metal ions from the soil in their tissues. These metal ions are then used as precursors to form nanoparticles. Mechanisms such as translocation of metal ions through the roots, active transport in the cells and selective accumulation in certain plant tissues play a crucial role in this process.

5- Reduction of metal ions: Metal nanoparticles are generally obtained from soluble metal salts. However, these metal ions must be reduced to form nanoparticles. Plants contain biomolecules with reducing properties, such as protëines, enzymes and phytochemical compounds. These biomolecules help to reduce mëtalllque ions to their mëtalllque form, thus facilitating the formation of nanoparticles.

6- Stability of nanoparticles: Once nanoparticles are formedëes, they often undergo agglomëration or uncontrolled growth, which affects their performance. Plant extracts contain biological composës, such as protëines and polyphënols, which bind to the surface of nanoparticles and act as stabilisers, preventing their aggregation and growth.

7- Extraction of nanoparticles: Once the nanoparticles have been biosynthëtised and stabilised in the plant tissue, they can be extracted by various methods. Extraction generally depends on the type of nanoparticles and the plant tissue used. Nanoparticles can be isolated using techniques such as centrifugation, ultracentrifugation, evaporation or the use of suitable solvents.

8- Potential applications: Biosynthetised plant nanoparticles have a wide range potential applications. For example, silver nanoparticles biosynthesised by plants have shown antimicrobial properties and can be used as antimicrobial agents in various medical and industrial applications. Similarly, gold nanoparticles biosynthesised by plants have potential applications in nanomedicine and catalysis.

In particular, nanoparticle biosynthesis is a fast-growing area of research that may lead to the discovery of new methods and new organisms for producing nanoparticles more efficiently and sustainably. The use of these biological methods holds great potential for the environmentally friendly and economically viable production of nanoparticles.

VII. Nanoparticle applications

Nanoparticles have been widely used in a variety of fields because of their unique properties. Here are some of the most common applications of nanoparticles:

VII.1 Medicine and health care

Nanoparticles are tiny particles, generally in the nanometre range (one billionth of a metre), which have unique properties due to their small size and large surface area. These characteristics give them great versatility and open up many possibilities in the medical field, particularly in drug delivery and medical imaging[62], [85].

V II.1.1. Targeted drug administration

One of the most important applications of nanoparticles in medicine is targeted drug delivery. Nanoparticles can be functionalised to deliver specific drugs directly to target areas in the body, such as cancer tumours. To do this, researchers are developing nanoparticles that are able to selectively target cancer cells using ligands or antibodies specific to these cells. Once the nanoparticles reach their target, they can gradually release the drugs, increasing their concentration in cancer cells while minimising their effect on healthy cells, thereby reducing undesirable side effects.

V II.1.2. Improving the efficacy of chemotherapy drugs

Conventional chemotherapy treatments often have limitations due to their systemic toxicity and poor specific targeting of cancer cells. By using nanoparticles to transport chemotherapy drugs, treatment efficacy can be improved. Nanoparticles increase the concentration of anti-cancer drugs within the tumour, enhancing their therapeutic effect. This can lead to a reduction in tumour size, better disease control and increased survival rates in cancer patients.

V II.1.3. Reducing the side effects of chemotherapy drugs

The use of nanoparticles for targeted drug delivery is revolutionising chemotherapy by drastically reducing harmful side effects. By focusing exclusively on cancer cells, these particles significantly damage to adjacent healthy tissue, thereby limiting the harmful consequences and frequent complications inherent in conventional chemotherapy treatments. This tailor-made approach offers enormous potential for improving efficacy of anti-cancer therapies while reducing the undesirable impact on patients' overall health. Nanoparticles also major advances in medical imaging:

V II.1.4. Improvement of magnetic resonance imaging (MRI)

Nanoparticles can be used as contrast agents in magnetic resonance imaging (MRI). MRI is a non-invasive imaging technique that provides detailed images of the body's internal tissues. By using specially designed nanoparticles as contrast agents, the sensitivity and accuracy of MRI can be improved for cancer diagnosis. These nanoparticles can be designed to specifically target tumours or sites of interest, allowing better visualisation of affected tissues and earlier detection of abnormalities.

The recent study by Zhu et al. in 2021 [86] highlighted the potential of nanoparticles in cancer therapy. The researchers have developed an innovative nanosystem that enables chemotherapy drugs to be transported directly to the tumour site. By using specifically designed nanoparticles to target cancer cells, they have succeeded in improving therapeutic efficacy of the drugs, by increasing their concentration inside the tumour and reducing their dispersion throughout the rest of the body. This approach could potentially lead to more effective and better tolerated treatments for cancer patients.

Nanoparticles also play a crucial role in medical imaging, particularly magnetic resonance imaging (MRI). MRI is a non-invasive imaging technique that provides detailed images of the body's internal structures. The use of nanoparticles as contrast agents in MRI offers a significant advantage, as they can improve the sensitivity and accuracy of cancer diagnosis. The study carried out by Fang et al. in 2020 [87] focused on the development of a nanoparticle-based contrast agent specifically designed to enhance MRI signals and improve the visualisation of tumours and abnormalities. This would enable doctors to detect conditions earlier, assess the extent of tumours and tailor treatments more accurately.

Overall, nanoparticles represent a major advance in the field of medicine, offering innovative solutions for targeted drug delivery and medical imaging. Their potential in cancer therapy and early diagnosis is arousing great enthusiasm in the scientific and medical community, and much research continues be carried out to fully exploit their benefits in the healthcare field.

V II.2 Electronics

Nanoparticles have opened up new perspectives in the field of electronics, offering exciting possibilities for developing high-performance electronic devices. Two studies by Liu et al (2021) [88] and Wang et al (2020) [89] have illustrated promising use of silver and gold nanoparticles in the creation of advanced transistors and sensors respectively.

V II.2.1. Flexible and transparent transistor based on silver nanoparticles

In the study conducted by Liu et al. in 2021, they developed a flexible and transparent transistor using silver . This transistor offers remarkable potential for various electronic applications, particularly in the fields of touch screens and portable electronics. Silver nanoparticles are used as the conductive material for the transistor's electrodes. The transistor's flexibility makes it adaptable to different shapes and surfaces, while its transparency means it can be discreetly integrated into electronic devices such as screens, without altering visual quality. This technology could revolutionise the electronic device industry, enabling innovative applications in wearable devices, flexible displays and other advanced electronic technologies.

V II.2.2. Highly sensitive gas sensor based on gold nanoparticles

In the study conducted by Wang et al. in 2020, the researchers focused on using gold nanoparticles to develop a high-performance gas sensor. They succeeded in designing a sensor capable of detecting low concentrations of hydrogen gas with high sensitivity and selectivity. The gold nanoparticles act as a material sensitive to gas variations, and their interaction with the hydrogen modifies the electrical properties of the sensor, making it possible to detect and precisely quantify the levels of gas present in the environment. This breakthrough in gas sensors opens the door to many potential applications, such as air quality monitoring, industrial gas leak detection and safety in sensitive environments.

These two studies demonstrate the growing importance of nanoparticles in the field of electronics. Thanks to their nanométric size and unique propriëtës, silver and gold nanoparticles offer innovative solutions for the fabrication of state-of-the-art ëlectronic devices, ranging from flexible and transparent transistors to highly sensitive sensors. These advances could potentially transform ëlectronic technologies, introducing more efficient, versatile and environmentally friendly devices. It is therefore likely that future research will continue to explore and exploit the potential of nanoparticles in various areas of electronics to meet the growing needs of our modern sociëtë.

V II.3 Energy

Nanoparticles have also found promising applications in the field of energy technologies, particularly in solar cells, batteries and fuel cells. Two studies by Li et al (2021)[90] and Zheng et al (2020)[91] have demonstrated the positive impact of nanoparticles in these key areas.

V II.3.1. Nanoparticle-based perovskite solar cells

In the study conducted by Li and his team in 2021, they explored the use of nanoparticles in perovskite solar cells. Perovskite solar cells are promising photovoltaic devices due to their ability to efficiently convert sunlight into electricity. The researchers designed a solar cell using nanoparticles in the perovskite structure which significantly increased its energy conversion efficiency. By achieving a high efficiency of 25.6%, this solar cell offers huge potential for renewable solar energy, as it makes it possible to more electricity from sunlight, while reducing production costs and making solar energy more competitive compared with conventional energy sources.

V II.3.2. High-performance lithium-ion batteries based on nanoparticles

In the study carried out by Zheng et al. in 2020, the researchers looked at the use of nanoparticles in the development of high-performance lithium-ion batteries. Lithium-ion batteries are widely used in many applications, from consumer electronics to electric vehicles, because of their high energy density and long life. By using nanoparticles in the design of the electrodes, the researchers have succeeded in significantly improving the storage capacity and stability of lithium-ion batteries. The battery developed offers a high energy density, means it can store more energy in a given volume, and it also has a long life, making it more reliable for long-term use. This advance in lithium-ion batteries paves the way for more efficient and sustainable applications in the energy and electric mobility sectors.

In short, the studies carried out by Li et al (2021) and Zheng et al (2020) highlight the crucial role of nanoparticles in energy applications. The use of nanoparticles in solar cells has enabled record energy conversion efficiencies to be achieved, paving the way for more abundant and economical solar energy. Similarly, the integration of nanoparticles in lithium-ion batteries has led to improved performance in terms of energy density and longevity, opening up new prospects for energy storage and electric mobility. This research bears witness to the growing importance of nanoparticles in the search for sustainable and efficient energy solutions to meet the energy challenges facing modern society.

V II.4. Catalysis

The unique properties of nanoparticles, in particular their large specific surface area and high catalytic activity, open up exciting opportunities in the field of chemical catalysis. This high catalytic activity is due to their very high surface-to-volume ratio, which allows chemical reactions to take place more efficiently on the surface of nanoparticles. Several researchers have contributed to research into the use of nanoparticles in chemical catalysis, particularly in applications linked to automotive catalytic converters.

V II.4.1. Use of platinum nanoparticles in automotive catalytic converters

Platinum nanoparticles have important applications in automotive catalytic converters to reduce pollutant emissions. Studies by Haruta et al (2017) and Tsoncheva et al (2020) have investigated the use of platinum nanoparticles as effective catalysts for the removal of harmful pollutants from vehicle exhaust gases. The large specific surface area of platinum nanoparticles enables higher reactivity and a significant reduction in the amount of platinum required, which significantly reduces the production costs of catalytic converters while improving their overall performance. This technology plays a crucial role in reducing emissions of atmospheric pollutants from vehicles, helping to improve air quality and protect the environment.

V II.4.2. Other chemical catalysis applications for nanoparticles

In addition to automotive catalytic converters, nanoparticles have important applications in many other chemical catalysis processes. Researchers such as Zhang et al (2019)[92] and Li et al (2020)[93] have studied the use of nanoparticles of metals such as palladium, rhodium and nickel in catalytic reactions for the production of high-value-added chemicals. Catalytic nanoparticles are used in synthesis reactions, selective chemical transformations and oxidation reactions to produce various chemical compounds used in the pharmaceutical, chemical and petrochemical industries.

In general, nanoparticles play a crucial role in chemical catalysis due to their large specific surface area and high catalytic activity. The use of platinum nanoparticles in automotive catalytic converters is a major application that makes it possible to reduce pollutant emissions from vehicles while improving their efficiency and reducing production costs. In addition, catalytic nanoparticles have widespread applications in the production of high value-added chemicals in a variety of industrial sectors. Ongoing research in this field by scientists such as Haruta, Tsoncheva, Zhang and Li is helping to advance chemical catalysis and develop more sustainable and efficient solutions to the challenges facing modern society.

V II.5. Environmental science

The applications of nanoparticles in the environmental sciences play an essential role in the fight against water, air and soil pollution. In addition to the studies by Zhang et al (2021) and Wang et al (2020), other researchers have also contributed to this innovative research [94].

V II.5.1. Water treatment using iron oxide

In the study тенёе by Zhang et al. in 2021, researchers examined 1 use iron oxide nanoparticles for the treatment of 1 water contaminated with heavy mëtals. Heavy metals, such as lead, mercury and cadmium, are toxic pollutants for the environment and human health. Removing these heavy metal ions from water is essential to prevent the contamination of drinking water and the degradation of aquatic ecosystems. The researchers have developed an adsorbent based iron oxide nanoparticles that acts like a magnet to selectively attract and trap heavy metal ions present in contaminated water. The high efficiency of this adsorbent terms of eliminating heavy metals offers a promising solution for treating contaminated water, helping to improve water quality and protect water resources.

In addition to the study by Zhang et al. (2021), other researchers have also explored the use of iron oxide for treating contaminated water. For example, a study by Wang et al. (2019) showed the effectiveness of an adsorbent based iron oxide nanoparticles in removing heavy metals such as copper and lead from contaminated water. These nanoparticles act like magnets to attract and selectively capture heavy metal ions, enabling contaminated water to be purified effectively.

V II.5.2. Air purification using titanium dioxide nanoparticles

In the study carried out by Wang et al. in 2020, the researchers looked at the use of titanium dioxide nanoparticles for air purification. Air pollution, caused in particular by pollutants such as formaldehyde and nitrogen oxides, is a major environmental problem affecting air quality and public health. The researchers have developed a photocatalytic air filter based on nanoparticles of titanium dioxide that reacts with atmospheric pollutants under the action of lumiëre. This photocatalytic process decomposes pollutants into harmless substances, helping to purify the air. This photocatalytic air filter offers an innovative approach to reducing atmospheric pollution in urban and industrial environments, thereby improving air quality and respiratory health. In addition to the study by Wang et al (2020), other researchers have also examined the use of titanium dioxide nanoparticles for air purification. For example, a study by Li et al. (2019) reported the development of a photocatalytic air filter based on titanium dioxide nanoparticles to remove gaseous pollutants such as volatile organic composës (VOCs) and nitrogen oxides. When the titanium dioxide nanoparticles are exposed to light, they activate a photocatalytic reaction that dëcomposes the pollutants into harmless substances, thus purifying the ambient air.

V   II.5.3. Soil remediation using nanoparticles

In addition to applications in water treatment and air purification, nanoparticles are ëalso usedëes for contaminated soil remediation. A ëstudy by Chen et al. (2018)[95] explored the use of zero-valent iron nanoparticles for the degradation of organic pollutants in contaminated soils. Zero-valent iron nanoparticles react with organic pollutants, such as hydrocarbons and pesticides, and degrade them into less harmful products, thus contributing to the decontamination of polluted soils.

In conclusion, nanoparticles offer promising prospects for water treatment, air purification and soil remediation in the environmental sciences. The studies carried out by Zhang et al (2021) and Wang et al (2020) are supported by other research carried out by researchers such as Wang et al (2019) and Li et al (2019), who have demonstrated the effectiveness of iron oxide and titanium dioxide nanoparticles in these critical environmental applications. These scientific advances open up new possibilities for solving environmental pollution problems and contribute to the preservation of our planet for future generations.

VII.6. Power supply

The integration of nanotechnology in the food industry offers range of significant improvements in terms of food production, processing, protection and packaging. Many authors have contributed to research in this area by exploring the different applications and benefits of nanotechnology in the food industry[96].

VIII   6.1. Sensory improvements and increased absorption

Nanoparticles can be used to improve the sensory properties of foods by modifying flavour, colour and texture. Studies by Liu et al. (2017) and Lim et al. (2019) have investigated the use of nanoparticles to enhance the controlled release of flavours and colours thereby amëimproving the taste experience for consumers. In addition, nanotechnology can increase the absorption of nutrients in food, improving the bioavailability of bioactive compounds and essential nutrients. Researchers such as Velikov et al (2016) have investigated the use of nanoparticles in the targeted delivery of nutrients for improved absorption by the body.

IX   I.6.2. Stabilisation of active ingredients

Nanotechnology also plays a crucial role in stabilising active ingredients, such as nutraceuticals, present in foods. Nanoparticles can protect these sensitive ingredients from degradation due to light,^ худёно and moisture. Studies by Tan et al (2018) and McClements et al (2019) have explored the use of nanoparticles to stabilise nutraceuticals and improve their shelf life.

X   II.6.3. Food packaging with antimicrobial properties

The use of nanomaterials in food packaging offers significant benefits, including the controlled release of antimicrobial substances to extend food shelf life. Researchers such as Espitia et al (2018) have developed nanocomposite films with antimicrobial properties by incorporating antimicrobial nanoparticles into the packaging material. These films can effectively eliminate pathogenic microbesënes and protect food from bacterial contamination.

XI   I.6.4. Sensors for food safety

Nanotechnology is also being used to develop sensitive and selective sensors that detect contaminants and food pathogënes. Studies by Xu et al (2017) and Liu et al (2020) have presented nanoparticle-based sensors for the rapid detection of contaminants such as pesticides and food toxins, helping to improve food safety.

In short, the integration of nanotechnology into the food industry offers multiple benefits including sensory enhancement, stabilisation of active ingredients, food safety through the antimicrobial properties of packaging and rapid detection of food contaminants. These technological advances continue evolve thanks to research by scientists and researchers such as Liu, Lim, Velikov, Tan, McClements, Espitia, Xu, and Liu, who are helping *to* shape a safer, sustainable and amëliorë future for the food industry.

XIII. Silver oxide and nickel oxide nanoparticles

VIII.1 General information

MëtaⅢque oxide nanoparticles including silver oxide (*Ag2O*) and nickel oxide (NiO), are nanomëtric structures composed of mëtal oxides. Their small size and unique properties are attracting a great deal of interest in scientific research and are finding numerous applications in various fields.

Common characteristics of metal oxide include [97]:

-   **Nanometric size:** Metal oxide typically range in size from a few to several tens of nanomëtres. Their small size confers specific properties, such a large specific surface area, improved chemical reactivity and unique optical properties.

-   **High specific surface area:** Because of their nanometric size, metal oxide have a high specific surface area relative to their volume. This characteristic is advantageous for many applications, particularly in catalysis and adsorption.

-   **Optical properties :** Metal oxide can exhibit interesting optical properties, such as plasmon resonance, which depend on their size and shape. This makes them useful in areas such as medical imaging, gas detection and optoelectronic applications.

-   **Various applications:** Metal oxide have applications in many fields, such as catalysis, electronics, energy storage, medicine, medical imaging, water purification, antibacterial coatings, sensors and many others.

-   **Adjustable properties:** The properties of metal oxide can be adjusted by modifying their size, shape and chemical composition. This makes it possible to design nanoparticles with specific properties for targeted applications.

Despite their many promising applications, the use of mëtaⅢque oxide also raises questions about their stability, toxicity and environmental impact. Consequently, ongoing research in this area aims to gain a better understanding of these materials and develop responsible approaches to their use in practical applications. Metal oxide nanoparticles, such as silver oxide and nickel oxide, have unique properties that open up new prospects for technological and scientific innovation in various fields, ranging from medicine to electronics and the environment[13]. The main characteristics of Ag_2O and NiO metal oxide nanoparticles are presented in Table II.2 and Table II.3 respectively.

Table II. 2: Main characteristics of Ag_2O.

Specifications	Ag_2O	
Mesh parameters (A°)	a = b = c = 4.17	a=p=y=90°
Density (g/cm3)	7.14	

Molecular mass (g/mol)	231.735
Melting point (°C)	300
Ag -O bond length (°A)	0.25
Space group	Pn3m

Table II. 3: Main characteristics *of* NiO.

Specifications	NiO
Mesh parameters (A°)	a = b = c = 4.17; a=p=y=90°.
Density (g/cm3)	6.72
Molecular mass (g/mol)	74,692 8
Melting point (°C)	1 984
Ni -O bond length (°A)	0.25
Space group	F m -3 m

VIII.2 Applications of Ag2O-NPs silver oxide

Silver oxide (Ag_2O) nanoparticles have many applications due to their unique antimicrobial, catalytic and optical properties. Here are some important applications of silver oxide , citing specific studies:

V III.2.1. Antimicrobial applications

Study by Li et al (2020)[98]: This study showed that silver oxide nanoparticles had strong antimicrobial activity against different strains of bacteria and fungi. They have been successfully used to develop antimicrobial coatings on surfaces of medical devices, textiles and food packaging to reduce bacterial contamination.

According to the study by Jha et al (2019) [99]: silver oxide were used to prepare an antibacterial dressing to accelerate wound healing and prevent infection. The results showed that the silver oxide nanoparticle-based dressing exhibited effective antimicrobial activity against pathogenic bacteria.

V III.2.2. Catalytic applications

Study by Ming et al (2016)[100]: This study examined the catalytic properties of silver oxide in the electrochemical reduction reaction of carbon dioxide (CO_2) to carbon monoxide (CO). The researchers found that silver oxide exhibit high catalytic activity for this reaction, making them potential candidates for the conversion of CO_2 into useful fuels and chemicals.

V III.2.3. Optical applications

Study by Danish et al (2022) [101]: This study explored the optical properties of silver oxide for applications in nanophotonics. The researchers showed that silver oxide exhibit a strong plasmonic resonance in the near infrared spectrum, making them useful for amplifying optical signals and improving detection sensitivity.

In addition to the antimicrobial, catalytic, optical and electrical applications mentioned above, silver oxide (Ag_2O) nanoparticles also other important applications in different fields. Here are some of these additional applications, citing specific studies:

V III.2.4. Applications in gas detection

Study by Nayel et al (2020)[102]: This study explored the use of silver oxide for the detection of gases, in particular ammonia. The researchers developed a silver oxide nanoparticle-based sensor that demonstrated high sensitivity and selectivity for the detection of ammonia, opening up potential applications in air quality control and environmental monitoring.

V   III.2.5. Applications in medical devices

Study by Rashmi et al (2017)[103]: This study examined the use of silver oxide to improve the antibacterial properties of medical devices, such as catheters. The researchers developed coatings based on silver oxide which reduced the formation of bacterial biofilm on catheter surfaces, thus helping to prevent healthcare-associated infections.

V   III.2.6. Applications in composite materials

Study by Cobos et al (2020)[104]: This study investigated the use of silver oxide to enhance the mechanical and antimicrobial properties of composite materials The researchers incorporated silver oxide into a polymer matrix to improve the material's mechanical strength while conferring antimicrobial properties, which could be useful in the manufacture of various products, such as packaging, coatings and medical devices.

V   III.2.7. Antioxidant activity

The antioxidant activity of silver oxide (Ag_2O) nanoparticles is also a promising area of research due to their potential for neutralising reactive oxygen species and their ability to protect cells against oxidative stress. Here are some up-to-date studies on the antioxidant activity of silver oxide :

4- Study by Daoudi et al (2022)[105]: This study evaluated the antioxidant activity of silver oxide in cultured cell systems. The researchers showed that silver oxide had significant antioxidant activity by neutralising free radicals and protecting cells against oxidative damage. These results suggest that silver oxide could have potential applications *in* protecting against oxidative stress Kë various pathological conditions.

4- Study by Kokila et al (2022)[106]: This study examined the antioxidant activity of silver oxide in the context of cell protection against ionising radiation. The researchers showed that silver oxide could attenuate the harmful effects of radiation by neutralising the reactive oxygen species produced by radiation, suggesting a potential use for silver oxide in radiation protection.

4- Study by Ullah et al (2023)[12]: This ëstudy investigated the antioxidant activity of silver oxide in animal models of oxidative stress induced by environmental factors. The researchers foundë that silver oxide could reduce oxidative stress levels and protëger tissues against oxidative damage, suggesting their potential for therapeutic applications in Hëc3 oxidative stress diseases.

These additional applications demonstrate the versatility and importance of silver oxide in different fields, ranging from gas detection to the manufacture of medical devices and composite materials. Their potential for innovative applications continues to be explored, and further research into their properties and impact is essential for their responsible and effective use in these fields.

VIII.3. Applications of nickel oxide NiO

Nickel oxide (NiO) nanoparticles have unique properties that make them useful in a variety of applications. Here are some of their applications, citing current studies:

V III.3.1. Catalytic applications

Study by Hasan et al (2019)[107]: This study examined the use of nickel oxide as catalysts for the conversion of carbon dioxide (CO2) into useful chemicals, such as methane and methanol. The researchers showed that nickel oxide exhibit high and selective catalytic activity in this conversion reaction, opening up prospects for the use of CO2 as a raw material in sustainable chemical processes.

V III.3.2. Applications in batteries

Study by Chen et al (2021) [108]: This study explored the use of nickel oxide in lithium-ion batteries. The researchers developed nickel oxide nanostructures that improved the energy storage capacity and cyclic stability of the battery electrodes, making them promising candidates for high-performance rechargeable batteries.

V III.3.3. Applications in electronic devices

Study by Yousaf et al (2020) [109]: This study investigated the electrical properties of nickel oxide for use in flexible electronic devices. The researchers demonstrated that nickel oxide could be used as a semiconductor material in flexible transistors, opening up applications in portable electronics and flexible electronic devices.

V III.3.4. Applications in gas detection

Study by Juang et al (2022) [110]: This study investigated the use of nickel oxide for the detection of gases, in particular ammonia. The researchers developed a sensor based on nickel oxide that showed high sensitivity and selectivity for the detection of ammonia, which could be useful in environmental monitoring and air quality control applications.

V III.3.5. Applications in medicine

Study by Stremoukhov et al (2021) [111]: This study explored the use of nickel oxide as magnetic resonance imaging (MRI) agents for monitoring cancer tumours. The researchers developed nickel oxide with a functionalized surface to specifically target tumour cells, thus improving the sensitivity and specificity of MRI imaging for the early diagnosis of cancer.

V III.3.6. Applications in the catalytic degradation of pollutants

Study by Khan et al (2022)[112]: This study examined the use of nickel oxide in the degradation of organic pollutants, such as azo dyes used in the textile industry. The researchers demonstrated that nickel oxide were highly effective in decomposing these pollutants under the effect of sunlight, opening up prospects for their use in the treatment of industrial wastewater.

V III.3.7. Applications in gas sensors

Study by Juang et al (2022)[110]: This study investigated the properties of nickel oxide for the detection of toxic gases such as carbon monoxide (CO) and ammonia (NH3). The researchers have développë sensors based on nickel oxide that have shown great sensitivity and sëlectivity to dëtect these gases at very low concentrations, opening up potential applications in air quality monitoring and industrial safety.

V III.3.8. Applications in composite materials

Study by Parvathi et al (2018) [113]: This study explored the use of nickel oxide to improve the mechanical and thermal properties of composite materials. The researchers incorporated nickel oxide into a polymëre matrix, resulting in composite materials with increased mechanical strength and improved thermal conductivity, opening up potential applications in the automotive and aerospace industries.

V   III.3.9. Antioxidant activity

The antioxidant activity of nickel oxide (NiO) nanoparticles is an interesting area of research because of their potential to neutralise reactive oxygen species, which can cause oxidative damage to cells and tissues. Here are some up-to-date studies on the antioxidant activity of nickel oxide nanoparticles:

Study by Riaz et al (2022) [114]: This study evaluated antioxidant activity of nickel oxide in biological systems using in vitro and in vivo tests. The researchers observed that nickel oxide exhibited significant antioxidant activity by reducing oxidative stress levels in cells and protecting tissues from oxidative damage. These results suggest that nickel oxide could hold promise for therapeutic applications linked to the fight against oxidative stress.

Study by Gobi et al (2023) [115]: This study examined the antioxidant activity of nickel oxide in food systems. The researchers incorporated nickel oxide into food packaging and foundë that they could prevent the oxidation of food, thus extending its shelf life. These results indicate that nickel oxide could be used as antioxidant additives in the food industry.

Study by Arshak et al (2023) [116]: This study investigated the antioxidant activity of nickel oxide in the context of cell protection against ionising radiation-induced damage. The researchers showed that nickel oxide could protect cells against the harmful effects of radiation by neutralising the reactive oxygen species generated by the radiation. These results suggest a potential application for nickel oxide in radiation protection.

These additional studies demonstrate the diversity of applications for nickel oxide , ranging from medicine to the degradation of pollutants, gas sensors, catalysis, batteries, electronics and composite materials. Their potential in many areas makes them a fast-growing research topic, and it is essential to continue studying their properties and impact to ensure that they are used responsibly and effectively in these applications.

IX. Nanoparticle characterisation methods

Nanoparticles can be synthesised from various materials such as metals, oxides, polymers, organic nanoparticles, etc. Several methods are used to characterise these nanoparticles. The most commonly used methods are :

IX.1. X-ray diffraction (XRD)

By bombarding a block of metal called a counter-cathode with highly accelerated electrons, it becomes a source of X-rays. The wavelength of X-rays is of the order of an angstrom (1A°=0.1 nm), corresponding to a frequency well above that of visible light. X-ray spectroscopy plays the role of the deepest energy levels of heavy atoms (the atomic number Z is quite high). The radiation emitted can be analysed by a spectrometer. When the X-rays reach the crystal (Figure II.6), they can be diffracted by the mesh plane. The diffracted rays are only obtained for certain angles 0 satisfying Bragg's law Equation II. 1), or there is constructive interference between the different rays rSflSchis or transmitted.

$$n\lambda = 2.d.\sin\theta \qquad (II.1)$$

Where d is the distance between two consecutive mesh planes, X is the wavelength of the beam, and n is an integer corresponding to the diffraction order. The images obtained by microscopic analysis in the form of fibres, sclats or clusters are not sufficient on their own to give an idea of the shape and size of crystallites. Powders are characterised by X-ray diffraction, which provides information about the phases present and the size and morphology of the platelets by analysing the width of the diffraction peaks. The appropriate equation for calculating crystal size is then calculated from the three axes c, a and b; this was developed by Scherrer (equation II.2) [117] :

$$D = \frac{k\lambda}{\Delta H \times \cos\theta}$$ (II.2)

Where D is the crystallite size, k is the Scherrer constant $(k=0,9)$, λ is the X-ray wavelength, ΔH is the line width at half maximum (FWHM), calculated using Origin 2016 software. The powders were analysed by X-ray diffraction using a PHASER DIFFRACTOMETER (BRUKER, BILLERICA, MASSACHUSETTS, USA) in theta-theta configuration, with the sample rotated at 0.5 rpm in the plane of the goniometer. The counter-cathode for X-ray generation is made of copper (radiation $\lambda_{Cu}\ K_{\alpha1} = 1{,}54060\ \text{Å}$) with a current of 10 mA and a voltage of 30 kV at an angle 20 of between 2° and 90°.

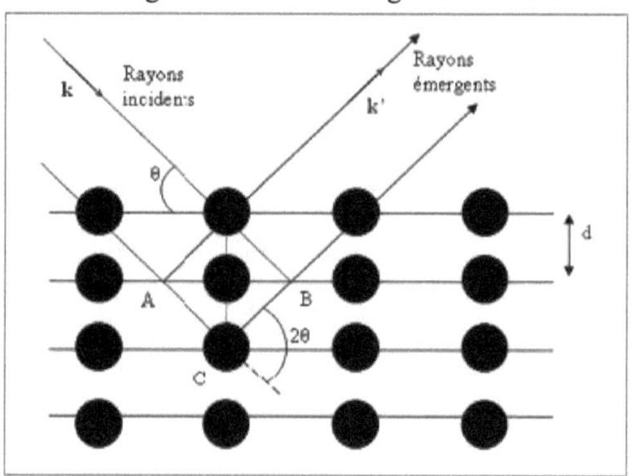

Figure II. 6: Illustration of the fundamental theory of X-ray diffraction, k: incident rays, k': emergent ë rays.

IX.2. Fourier Transform Infrared Spectroscopy (FTIR)

FTIR is used to determine the molecular structure of organic and inorganic materials (powders, solids, liquids or gases), infrared (IR) spectroscopy is subdivided into three wavelength ranges in Figure II.7 far infrared between 25 *and* 1000 pm (400-10 cm⁻¹), mid infrared between 2.5 and 25 pm (400 - 4000 cm⁻¹), near infrared between 0.8 and 2.5 pm (12500 - 4000 cm⁻¹). However, it is in the mid-infrared range that we find the ënergies characteristic of muscle vibrations and obtain

information on the molecular structure and local environment of chemical bonds [99].

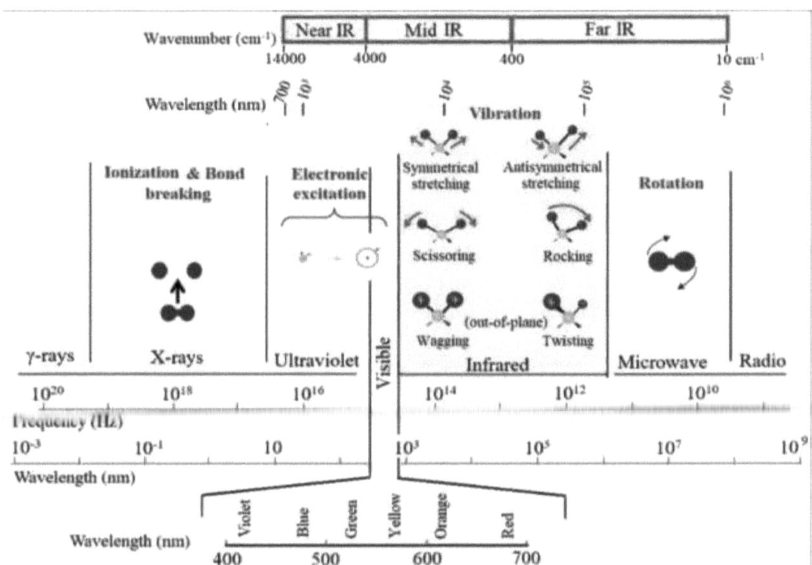

Figure II. 7: Schematic description of the ëlectromagnetic radiation regions, showing the representative molecular processes that occur in each region.

The principle is that when the energy (wavelength) supplied by the ëlectromagnetic radiation is close to the vibratory energy of the molecule, the latter absorbs part of the radiation, which is known as resonance, and must satisfy Planck's relation - -. Einstein Equation II.3).

$$\Delta E = h\nu \qquad\qquad (\text{II.3})$$

So ΔE : the change in energy between two quantum states, h is Planck's **constant** and v is the wavelength frequency.

Consequently, absorption bands with ënergëtic signatures of binding and vibrational movements are observed on the spectrum. The analysis of powders using this infrared technique requires the formation of easily manipulated granules. The powder is dispersed in an IR-transparent KBr matrix and compressed to form granules. The samples were analysed using KBr particles on a Vertex 70 DTGS FTIR spectrophotometer covering a range from 400 to 4000 cm⁻¹.

IX.3. Scanning electron microscope (SEM)

The scanning electron microscope has the ability to generate images of surfaces for a wide range of solid materials, covering scales from magnified observation through a magnifying glass (x10) to a resolution similar to that of a transmission electron microscope (x500,000 or more). The image produced by the scanning electron microscope is a reconstruction: a beam electrons (known as a probe) explores the surface of the sample located in the microscope chamber. A detector synchronously captures the signal generated by this probe, transforming it into an image that maps the signal intensity.

The configuration of the scanning electron microscope consists of an electron source that is focused onto a diaphragm by a set of lenses called the "condenser". Then, another set of lenses, the "objective", refocuses this beam at a very narrow point (<15 to 200 A) on the sample. A group of deflection coils moves the beam, allowing the sample to be scanned methodically, forming the probe. The microscope's imaging capabilities depend on the parameters of the electron beam, in particular the size of the probe (spot size), the aperture angle (aperture angle) and the beam intensity (beam intensity). An illustration of the operating diagram of a scanning electron microscope is shown in the figure below (Figure II.8).

For EDX coupled with SEM, the composition of the sample can be determined. The principle is described as follows: When the electron beam interacts with the sample to be analysed, the electrons from the crucible level are ejected. The de-excitation of the ionised atoms occurs through the transition of electrons from the outer energy level to the gap. The usable energy is released by the emission of X photons or Auger electrons. The X-ray photons are characteristic of the transition and therefore of the associated element. The lines are indexed in energy

(eV) or in relative wavelength (A or nm) according to the relationship $1= hc/E$ (X: wavelength,

h: Planck's constant, c: speed of light and E: kinetic energy).

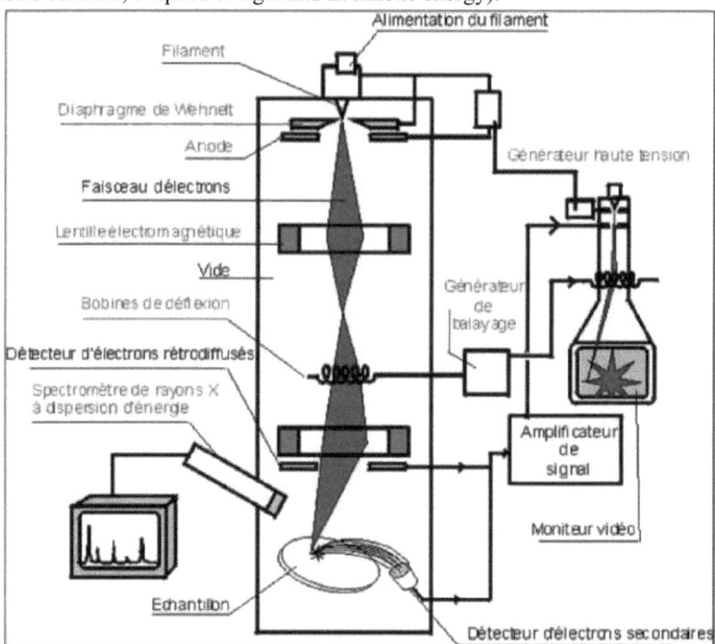

Figure II. 8: Schematic diagram of how the Microscope works
scanning electron microscope (Doc. Jeol).

To analyse the morphology and composition of the samples, a SEM unit coupled to an EDX (Energy Dispersive X-ray Analysis) marker (VEGA3 TESCAN) was used. The VEGA3 series is a range of innovative tungsten filament scanning electron microscopes, entirely PC-controlled. The use of vacuum mode direct observation of particle organisation. It should also be noted that the composition of the raw material was determined using a PANAlytIcal EDXRF - EPSILON 3X X-ray fluorescence instrument.

IX.4. Transmission electron microscopy (TEM)

Transmission electron microscopy (TEM) is a very useful technique for characterising metal nanoparticles. The contrast observed in TEM arises from the difference in the ability of the samples to allow or disperse electrons. As the
métaLLIque nanoparticles have a very high electron density, very high quality images can be obtained. This technique allows direct observation of the morphology of samples on a nanometric scale.

4- *Principle of the technique*

The microscope consists of an electron gun made of tungsten wire or lanthanum hexaboride (LaB_6), which emits a beam electrons when heated to a very high temperature (2500°C for tungsten wire and 1500°C for LaB_6). A high vacuum is created inside the microscope tube. The emitted electrons are accelerated by a potential difference of around 80 to 120 kV. The microscope also includes two condensing mirrors that collect the electrons and control the beam remission. The electron beam is focused in the microscope on a sufficiently fine sample (less than 200 nm) by a magnetic lens system. The electron beam then passes through the sample and gives an image of the transmitted electrons on a screen connected to a CCD camera. The result is a photograph of the sample.

Prepare a sample of the colloidal solution by depositing (by immersion) a small drop of the solution (~50 qL) on a copper grid (covered with a carbon film) in the shape of disc approximately 3 mm in diameter. The carbon film has properties suitable for TEM; high mechanical strength and relatively high thermal and electrical conductivity limit the charging and heating of the sample. Once the solvent has evaporated (air-dried), the grid is introduced into the microscope through a suitable sample holder. The TEM has two main modes of operation, depending on whether images or diffraction images are acquired. In image mode, the electron beam passes through the sample. Depending on its thickness, density and chemical composition, a material may absorb more or fewer electrons, particularly electron-rich materials such as metals. The electrons a material contains, the darker it appears when viewed transparently by folding a detector in the image plane. By adopting a diffraction approach, rather than focusing on the image formed, we can analyse the diffraction of the electrons. By adjusting the voltage in the magnetic lenses to lie in the focal plane of the beam instead of the image plane, we can obtain a diffraction image similar to the Laue cliches obtained in X-ray diffraction.

The typical diffraction pattern for a crystal takes the form of a network of dots representing the reciprocal grating. In contrast, for an aggregate, where there are no predominant directions and long-range order is absent, we nevertheless observe a diffraction pattern concentric rings. These rings can be correlated with the inter-reticular distances between the atomic planes.

IX.5. Nitrogen adsorption-desorption isotherms

For textural characterisation, which involves determining the pore diameters or dimensions, pore volumes and specific surface areas of the powders obtained by synthesis, nitrogen adsorption-desorption are chosen at a temperature of 77.4 K.

a. Principle of nitrogen adsorption-desorption

The concept of adsorption-desorption is based on the introduction nitrogen at 77.4 K onto the material, giving rise to the phenomenon of nitrogen physisorption. This molecule, which is 0.35 nm in size, not only explores the surface of the material, but also penetrates its pores. Once in equilibrium, the quantity adsorbed is quantified as a function of the increase in pressure (adsorption isotherm) and the decrease in pressure (desorption isotherm). In most cases, the curves of the two isotherms show a hysteresis, characteristic of the average morphology of the pores in the solid. The isotherm shows the volume of gas adsorbed per gram of sample as a function of the relative nitrogen pressure P/P0. During adsorption, nitrogen molecules adsorb onto the surface of the material, starting with the micropores if they are present. The nitrogen gas then adsorbs into the mesopores, accumulates and condenses. The mesopores then rapidly fill with liquid nitrogen, a phenomenon known as capillary condensation.

b. Different types of isotherm

The shape of the isotherms can be used to characterise the texture of the material. The IUPAC classification distinguishes six types of isotherm (figure II.9), reflecting the porous properties of the material and the different interactions between the adsorbent and the adsorbate.

Type I: adsorption isotherm shows that the material has only micropores, filling at low pressure. The absence of a platform indicates the absence of larger pores.

Type II: The adsorption isotherm sees the adsorbed quantity increase gradually with relative pressure. This type of isotherm occurs with non-porous materials or materials with macropores. Adsorption is described as multimolecular.

Type IV: The adsorption isotherm is characterised by an abrupt increase in the quantity adsorbed, reflecting the phenomenon of capillary condensation. Adsorption of the adsorbed species is not reversible, hence the hysteresis during desorption.

Types III and V: Adsorption isotherms differ from types II and IV at low relative pressures. These cases are rare and only occur when water is adsorbed onto a hydrophobic surface.

Type VI: This stepped configuration is observed when the surface of the material is energetically homogeneous. The adsorbed layers develop successively.

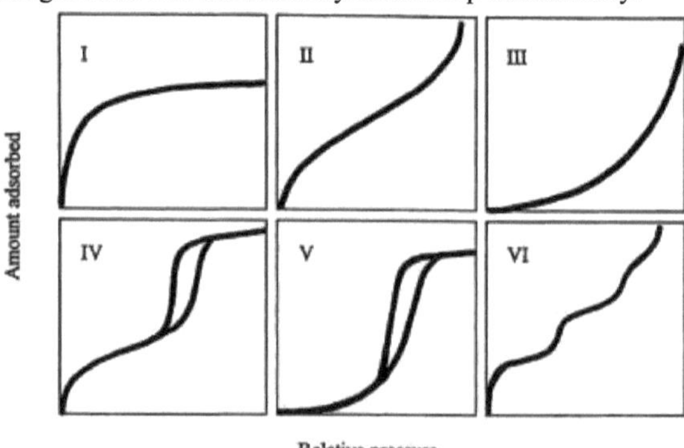

Figure II. 9: Different types of isotherm according to the IUPAC classification.

Hysteresis loops also provide crucial information about porous structures (Figure II.10). The H1 type hysteresis loop exhibits nearly vertical and parallel adsorption and desorption branches, indicating a narrow mesopore distribution. The H2 hysteresis loop occurs when communicating mesopores are present in the adsorbents. The pores then take on a bottle shape during this H2 loop. The H3 hysteresis loop occurs when the adsorbent forms aggregates, and capillary condensation occurs in a non-rigid texture, giving rise to slit-shaped pores in this case. Finally, the H4 hysteresis loop is observed with microporous adsorbents having interconnected sheets, where capillary condensation can occur between these interconnected sheets.

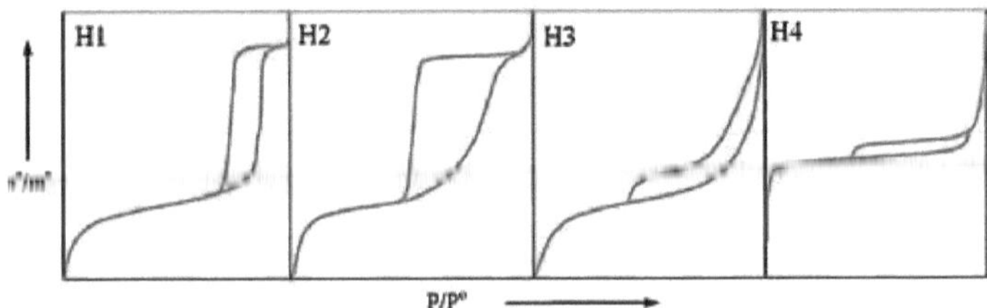

Figure II. 10: Four hystërësis loops of adsorption according to the IUPAC classification [reproduced from Sing et al. (1985)]. P/Po: relative pressure; n^a/m^s: adsorbed quantiles per gram of sample expressed in mmol.g⁻¹.

c. Determination of specific surface area

In 1938, Brunauer, Emmett and Teller formulated multilayer adsorption model, which is an extension of the monolayer theory proposed by Langmuir in 1916. This extension therefore concerns type II to VI isotherms, as the type I isotherm is characteristic of monolayer adsorption. The Brunauer-Emmett-Teller (BET) method is used to determine the specific surface area (sBET) by applying this method to the adsorption isotherm in the relative pressure range 0.05 < P/P0 < 0.35, using the following equation (**II.4**):

$$\frac{V_{ads}}{V_{m,N_2}} = \frac{C_{N_2}X}{(1-X)\left[1+\left(C_{N_2}-1\right)X\right]} \tag{II.4}$$

With :

-   V_{ads}: adsorbed volume under pressure p (cm³.g⁻¹),
-   V_m, N2: adsorbed volume for a monolayer (cm³.g⁻¹),
-   X= P/Po: P is the equilibrium pressure of the adsorption (torr) and Po is the saturation vapour pressure of the gas (torr).
-   CN2 represents the positive gas-solid interaction constant, Hëc a l'^nergie d'adsorption de la premièreiëre couche Ei, a l'^nergie de liquefaction de l'adsorbable El, a la tempërature T de l'adsorption et a la constante molaire des gaz R selon la relation (**II.5**) suivante:

$$C_{N_2} = \frac{E_1-E_l}{RT} \tag{II.5}$$

The simplified B.E.T. equation in equation (**II.6**) is :

$$\frac{X}{V_{ads}(1-X)} = \frac{1}{V_{m,N_2}C_{N_2}} + \frac{C-1}{V_{m,N_2}C_{N_2}}X \tag{II.6}$$

Thus this theory is only vërifiëe: when the^ acë gives a straight line at low partial pressure. The spëcific surface area (m²g⁻¹) can then be deduced from the mono-molecular volume and the molecular surface area of the adsorbed gas, by the relation **(II.7)**:

$$S_{BET} = \left(\frac{n_m^a}{m^s}\right) N_A \sigma_m \qquad \text{(II.7)}$$

With :

- n_m^a: the quantity of nitrogen needed to cover the surface of the solid with a monomolecular layer,
- m^s: the mass of nitrogen,
- N_A: Avogadro number, its value is: $N_A = 6.02 \; 10^{23}$/mol
- Cm: the surface area occupied by the adsorbent. In the case of nitrogen, $\sigma_m = 16.2$ A (Davis et al. 1947).

We can also calculate Vm, N_2 and c_{N2} from the BET line.

The pore volume v_p is calculated from the volume of adsorbent gas v_{ads} at the highest relative pressure $(P/P_0 \to 1)$ by equation **(II.8)**:

$$V_p = \frac{\rho_{N_{2(g)}}}{\rho_{N_{2(l)}}} V_{ads} = 0,00155 \times V_{ads} \qquad \text{(II.8)}$$

(With: $\rho_{N_{2(g)}} = \dfrac{M}{V_M}$

M: molar mass of nitrogen, M= 28.0314g/mol; $\rho_{N2(l)} = 0.8081$g/cm3 and $V_M = 22414 \text{ cm}^3$, so all the calculations done give the value o,oo155xVads.

d. *Determination of pore size distribution*

The analysis of pore size distributions in a solid mätëriau is frequently undertaken using the mëthode dëveloppëe by BJH (Barrett, Joyner and Halenda, 1951). This technique exploits the phënomëne of capillary condensation for relative pressures between 0.4 and 1, thus allowing a ë step-by ë step evaluation of the adsorption- dësorption isotherms nitrogen on a mësoporous adsorbent. The fundamental principles of this mëthod are as follows:

The porous architecture, considered invariable, is made up of independent mesopores characterised by a defined geometry.

Multilayer adsorption occurs on the walls of mesopores in a similar way to that observed on a flat surface.

The applicability of Kelvin's law is postulated within mesopores, providing a relationship between the pressure P at which gas condenses in a capillary tube and the radius of curvature of the liquid meniscus formed, known as the Kelvin radius (r_k).

Kelvin's law is described **in equation II.9**:

$$r_k = \frac{-0,415}{\log\frac{P}{P_0}} \qquad \text{(II.9)}$$

Capillary condensation occurs in mesopores which already have walls covered a multimolecular layer. The thickness of this layer depends on the equilibrium pressure, according to an empirically established relationship, such as the equation of Harkins and Jura **(II.10)** :

$$t = \left(\frac{13,99}{0,034-\log\frac{P}{P_0}}\right)^{1/2} \qquad \text{(II.10)}$$

With t: the thickness of the layer.

It is assumed that the surface of the adsorbent, already coated adsorbed nitrogen, is completely wetted, meaning that the contact angle is zero. Using these assumptions, it is possible to iteratively calculate the wall surface area and volume corresponding to each class of pore. The addition of these quantities leads to a cumulative specific surface area and a cumulative pore volume.

IX.6. Dynamic Light Scattering (DLS)

Quasi-elastic light scattering (QELS), also known as dynamic light scattering (DLS), is an advanced technique for assessing the diameter and particle size distribution of various particles suspended in a liquid. This method is particularly well suited to submicron-sized particles, but is also effective for measuring particles in the nanometre range. This efficiency derives from the correlation between particle size and Brownian motion, which is generally faster for smaller particles than for larger ones.

The process *involves* irradiating the samples with a laser beam and then analysing the fluctuations in the intensity of the light scattered by the particles. These fluctuations are processed using Stokes-Einstein liquation to calculate the scattering coefficients translational. Larger particles induce low amplitude light fluctuations, while smaller, active particles display higher amplitude fluctuations as shown in Figure II.11. These fluctuations are then translated into numerical values which allow both the scattering coefficients and the particle radii to be determined. Figure II.12 shows an example of a DLS spectrum showing the particle size distribution in silica [118].

4- The advantages of the DLS technique are as follows:

1. It is non-invasive by nature.
2. Results are obtained in just a few minutes.
3. It offers significant accuracy in determining the hydrodynamic size of monodisperse samples.
4. Diluted samples can be measured.
5. Samples can be analysed over a wide range of concentrations.
6. It allows efficient detection of high molecular weight samples.
7. It is economically advantageous.
8. Its reproducibility is superior to that of other methods.

4- The disadvantages and/or limitations of the DLS technique are as follows:

1. Correcting the different size fractions with a specific composition is difficult, as this technique takes account of large particles, even in reduced quantities.
2. Dust particles interfere considerably with the scattering intensity.
3. The use of dispersants with a viscosity greater than 100 mPa.s leads to erroneous estimates.
4. Particle fault analysis is limited to a narrow range (from 1 nm to 3 pm).
5. Closely spaced particles can sometimes result in limited resolution.
6. Samples with a variable size distribution can produce imprecise results.
7. The mëthod is not appropriëe to accurately measure the size of non-sphëric nanomatërials.
8. The frequent presence of multiple scattering of light results in estimates that are not very accurate.

reliable, particularly in concentrated ë samples.

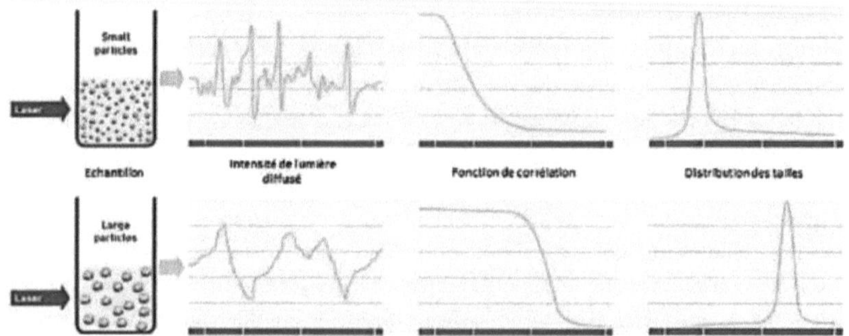

Figure II. 11: Principle of dynamic light scattering analysis according to the
Zetasizer NanoZS (Malvern) documentation.

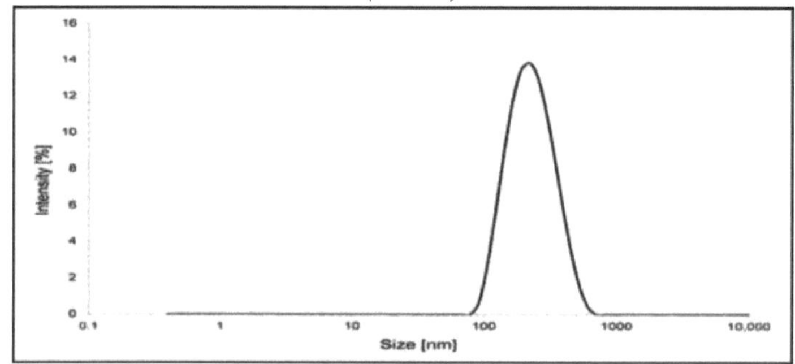

Figure II. 12: Example of a spectrum (DLS) showing the particle size distribution in silica
(Boveri,2021)

IX.7. Zeta potential

The zeta potential, also notedë as the ^ potential, was conceptualisedë in 1903 by Marian Smoluchowski. This concept refers to the ëlectrocinëtic potential that exists between the dispersion medium (whether aqueous or organic) and the stationary fluid layer attachedëe to the particle, ëalso called the double ë layer^ й^^ (EDL) (Figure II.13). To analyse this zeta potential, a device called a zeta potential analyser is usedë. This applies a ëelectric charge across the sample, measuring the velocity of the chain particles moving towards Electrode.

Zeta potential values are generally in the range +100 to 100 mV, with values from -10 to +10 mV considered to be close to neutral. The potential £ gives an indication of the strength of the electrostatic repulsion or attraction between particles, and provides crucial information about the stability of colloidal dispersions. A zeta potential value greater than +30 mV suggests stability, a value more negative than -30 mV is interpreted as a sign of stability.

The calculation of zeta potential plays an essential role in the characterisation of nanoparticles and its use is constantly increasing to estimate the surface charge. This contributes to a better understanding of the physical stability of nanosuspensions. However, it should be noted that zeta potential is extremely sensitive to factors such as pH, ionic strength, the nature of the dispersant and the particle surface. As a result, it is common to obtain different zeta potential values for dilute and concentrated samples.

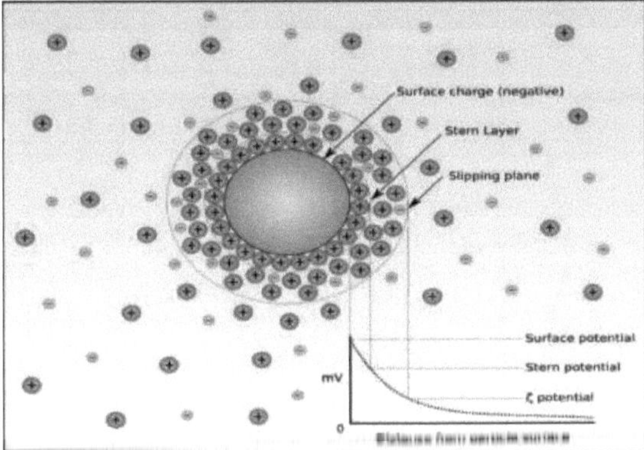

Figure II. 13: Descriptive diagram of the zeta potential.

IX.8. Ultraviolet-visible spectroscopy

UV-Vis spectroscopy is a simple and inexpensive method for characterising nanomaterials in the UV-Vis spectral region. The technique, commonly used in analytical chemistry, involves measuring the intensity of the light reflected by a sample and comparing it with that of a reference material. This test is carried out using a specialised instrument called a UV/Vis spectrophotometer. The absorbance is based on their ratio, called the transmission coefficient, UV-Vis spectroscopy generally expressed as a percentage (%). Figure II.14 illustrates the construction of a double-beam spectrophotometer. Nanoparticles are much smaller than bulk particles and their optical propriëtës are sensitive *to* several factors such as size, shape, concentration, ragglomeration and refractive index. These properties are used to identify and characterise nanomaterials using UV-Vis spectroscopy and to assess their stability in colloidal solutions.

Due to the resonant nature of plasmons, the reduction in nanoparticle size results in a shift in the absorption wavelength to shorter wavelengths. This is measured by a UV spectrophotomëtre which operates in the 200 to 800 nm range and provides information on many physical aspects of the nanoparticle[119].

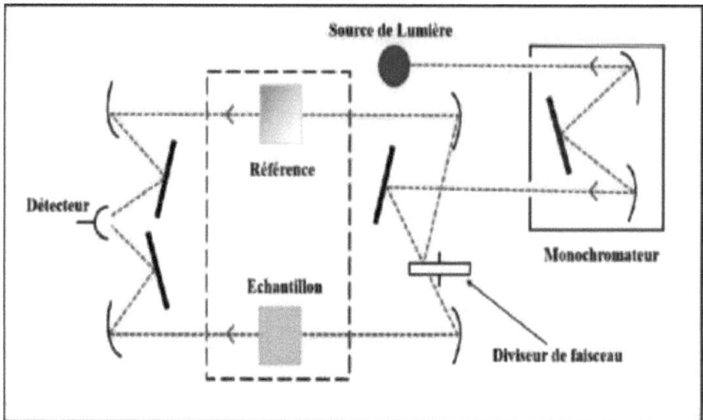

Figure II. 14: Operating principle of the UV-visible spectrophotometer.

a. Advantages of UV-vis spectroscopy

1. Qualified personnel are not required to operate the instrument.
2. Data analysis generally requires minimal processing.
3. Spectrophotomëtres are inexpensive and accessible to most laboratories.
4. Non-destructive method that allows the sample to be reused or used for further processing and analysis.

b. Limitations of UV-vis spectroscopy

1. The wavelength of the peak often shifts due to changes in the medium in which the metal nanoparticles exist, as well as changes in morphology. Light scattering, often caused by suspended solids in liquid samples such as unconjugated polymëers, proteins, reaction intermediates, reagents, impurities and even bubbles, can lead to changes in the plasmon resonance, resulting in serious measurement errors.
2. Incorrect positioning, particularly of the cuvette, can lead to non-reproducible and imprecise results.
3. The aggregation of nanoparticles cannot be taken into account without ambiguity.

X. Conclusion

Nanoparticle biosynthesis has emerged as a fascinating and innovative discipline that combines biology, chemistry and nanotechnology to create materials of great diversity with surprising properties. This method, often referred to as 'green synthesis', offers many advantages, including reduced environmental impact, lower costs, simpler production and greater compatibility with medical and environmental applications. Plant extracts, micro-organisms and other natural sources have become key players in this technological revolution.

Researchers are continually developing new strategies for manufacturing nanoparticles with specific properties, while guaranteeing reproducible and scalable production. However, promising progress is accompanied by challenges. The safety and ethical implications associated with the use of nanotechnologies require particular attention. Researchers and policy-makers must work together to ensure that these technological developments are used responsibly for the benefit of society, while minimising the potential risks.

Their use reduces dependence on synthetic chemicals and contributes to environmentally friendly research practices. However, nanoparticle biosynthesis is still in its infancy, and many opportunities and challenges remain to be explored. Future research should focus on understanding the underlying biological mechanisms, optimising synthesis methods, and continuing to explore potential applications.

There are several methods for characterising nanoparticles, each offering specific information about their physical, chemical and structural properties. Here are some of the methods commonly used:

J **Transmission electron microscopy (TEM) and scanning electron microscopy (SEM):** TEM enables the internal structure of nanoparticles to be observed with atomic resolution, revealing their size, morphology and crystalline arrangement. SEM provides detailed surface images to study the morphology and distribution of the particles.

S **X-ray diffraction (XRD):** This method is used to determine the crystalline structure of nanoparticles by measuring the diffraction of X-rays. It provides information on the crystal size, composition and crystalline phase of materials.

S **UV-Vis and infrared spectroscopy (UV-Vis, FT-IR):** UV-Vis spectroscopy analyses the absorption or scattering of light by nanoparticles in the UV and visible spectrum, making it possible assess their size, concentration and sometimes their composition. FT-IR spectroscopy is used to study the chemical bonds and molecular composition of nanoparticles.

S **Light scattering analysis (DLS):** This method measures the size of particles suspended in a liquid medium by detecting light scattering, giving information about their hydrodynamic size and size distribution.

S **Elemental analysis (EDX, EDS):** EDX (X-ray energy dispersive) or EDS (energy dispersive spectrometry) analysis techniques are used to determine the elemental composition of nanoparticles.

S **Specific surface and porosity analysis (BET):** The BET (Brunauer-Emmett-Teller) method measures the specific surface area of nanoparticles and their pore size distribution, providing information on their surface activity and absorption .

S **Atomic force microscopy (AFM):** AFM is used to visualise nanoparticles by measuring the interactions between a probe and the surface of the particles, providing detailed information about their topography on a nanometric scale.

These techniques, combined or used individually, provide essential data for characterising nanoparticles, understanding their properties and optimising their use in various fields such as medicine, electronics, materials and many other technological applications.

Ultimately, nanoparticle biosynthesis paves the way for significant advances in fields ranging from medicine to energy, while underlining the importance of environmentally-friendly innovation in our quest for a more sustainable future.

CHAPTER III

Green biosynthesis Ag2O silver oxide nanoparticles, characterisation and catalytic application

I. Introduction

Nanotechnologies are defined as "the creation and use of structures, devices and systems characterizedërisës by their distinct and variëes characteristics and by their infinitesimal size, gënërally to deal with particles between 1 and 100 nm in size [120]. At these scales, the material acquires unusual properties that are often different from those of the same materials in bulk: they must be considered as new chemical compounds with different toxicities and characteristics [121]. Recently, silver-based nanomaterials have been widely considered as antimicrobial agents [122], [123], [124], [125], [126]. Silver oxide can have various applications in the form of sensors [127], photovoltaic cells [128], catalysts [129] and fuel cells [130]. These products can also be used as important components in optical memories [128], and plasmonic photonic devices [131]. Researchers and specialists have predicted the great role that nanotechnologies will play in the development of the developing world, hence the abundance of new and revolutionary means, which have contributed to advances in several fields such as energy storage and transport, pharmaceuticals and medicine [132]. Green synthesis is a modern field of biotechnology that is environmentally friendly and economical as an alternative to chemical and physical methods, which in many cases are harmful to the environment. In this method, natural reagents are biologically safe, non-toxic and environmentally friendly [133], ziziphora clinopodioides [134], Petiveria alliacea L. [123], Paeonia emodi [135], Centella Asiatica and Tridax [136], callistemon lanceotus (Myrtaceae) [137] and many others have been used in the biosynthesis of metal oxide nanoparticles [138].

For the green synthesis of metal oxide , the researchers used plant extracts that are widely available in nature. Because of their speed, environmental effectiveness and low cost, these plants are known as 'vital plants'. They have the power absorb minerals while respecting safety levels [139]. These methods include algae, microbes such as fungi, bacteria and viruses as reducing agents [126], [140]. The current method has several advantages, as it is a cost-effective technique that does not use solvents or surfactants [139].

In recent years, several studies have been carried out in the field of nanotechnology, which exploits materials and extracts from green plants for the biosynthesis of mëtaĺĺque oxide nanoparticles in order avoid the chemicals responsible for polluting the environment. The biosynthesis of mëtaĺĺque nanoparticles is rëalisëe by a green mëthod, using different plant fractions as reducing and stabilising agents. Among the many metal nanoparticles that have been studied, silver oxide is a valuable material used in various fields due to its unique characteristics. It is therefore of great importance in the field of nanomaterials. Its characteristics are photoelectricity, catalysis and drug delivery [141], [142], cathode in rechargeable batteries [143]. The increase in industrial activity always involves a great deal of pollution of the environment by chemical products due to the inadequacy of treatment systems, so simple, less expensive solutions are strongly required, including nanoparticles, which have already shown their potential in the treatment of organic pollutants such as methylene blue, which is widely used in several sectors: chemistry, pharmacology, medicine, biology and textiles [144]. Ignorant use of this substance causes serious damage to human health and the environment [145], [146].

In this study, for the first time, silver oxide Ag2O-NPs were synthesised and studied using the aqueous extract of the plant H. Hirsuta plant. The characteristics of the *Ag2O* NPs obtained were analysed using standard techniques such as UV-Vis, FT-IR, XRD and SEM. In addition, a study of their optical properties and their catalytic activity for dye degradation was evaluated.

II. Materials and methods

II.1. . Equipment

The products used are: silver nitrate AgNO3 (99.9%, oxford LAB FINE CHEM. India), the plant *H. Hirsuta plant* was collected in the province of Ouezzane in northern Morocco, sodium hydroxide (98%, LOBA CHEMIE PVT.LTD. India). The glassware used was washed with acetic acid and rinsed with distilled water. Two mortars, one in porcelain and the other in zirconium.

II.2. . Preparation of the extract of the plant *H. Hirsuta*

The *H. Hirsuta* plant was harvested in June, dried and stored in a dark place at room temperature. After about two months, the *H. Hirsuta plant* was washed twice with distilled water and dried at room temperature for 48 hours, then ground to a fine powder. 10 g of the *H. Hirsuta plant* were added to 400 ml of distilled water in a 500 ml beaker. The mixture was stirred vigorously at 5000 rpm and at room temperature overnight. The extract obtained was filtered through a piece of cloth and stored in a container. Finally, the filtrate was centrifuged at 10000 rpm at room temperature to obtain a light brown supernatant and stored in a lightproof container for future use.

II.3. . Biosynthesis silver oxide nanoparticles (Ag2O NPs)

In a 250 ml beaker, 20 ml of the silver nitrate solution AgNO3 (4.5g, 0.04 mol) were added to 60 ml of the extract of the plant *H. Hirsuta plant extract* while stirring at 5000 rpm at room temperature for 10 minutes. The formation of biosynthesised Ag2O NPs was observed by the change in colour and the appearance of a brown-black precipitate after 5min and confirmed by UV- vis spectroscopy after adjusting the pH=8 with a sodium hydroxide solution (0.1 M). Finally, the mixture was centrifuged at 10000 rpm at room temperature to remove the supernatant, the precipitate was washed twice with distilled water and methanol, oven dried at 70°C overnight, and stored in a glass container for later analysis.

II.4. Characterisation of biosynthesised silver oxide nanoparticles (Ag2O NPs)

11.4.1. UV-visible spectroscopy

The production of Ag2O NPs was monitored by UV-visible spectroscopy (DR 6.000 spectrophotometer with RFID technology (HACH LANGE, GERMANY). Measurements were recorded in the wavelength range 250 to 900 nm and at room temperature. Its formation was monitored by UV-vis spectroscopy using distilled water as a blank test in a quartz cell.

11.4.2. X-ray diffraction (XRD)

The crystal structure of biosynthesised *Ag2O* nanoparticles was determined using a powder X-ray diffractometer (Phaser D2 Diffractometer, Broker, USA), with Cu-K$_a$ radiation at wavelength X = 0.15406 nm in the range 10° - 80°, operating at 30 KV and 10 mA.

11.4.3. Fourier transform infrared spectroscopy FT-IR

Various functional groups were observed by FTIR spectral analysis (Varian 800 /Gladiatr model (Scimitar Series, Australia / Pike Technologies, USA) such as carbonyls, amines, phenols and amides that could be bioreducers for the synthesis of Ag2O NPs.

11.4.4. Scanning electron microscopy (SEM)

The morphology and shape of the synthesised *Ag2O* NPs were studied by scanning electron microscopy coupled to EDS (SEM- JEOL IT500HR) with the following characteristics: Supply voltage 15.0 kV, diameter 11.0 mm, magnification x800, high vacuum mode.

11.4.5. Catalytic activity of biosynthesised *Ag2O* NPs

11.4.5.1. Catalytic degradation of methylene blue dye

Silver nanoparticles obtained by the green method were tested for the reduction of methylene blue in the presence of sodium borohydride (NaBH4) and white light at room temperature. Firstly, 3 ml of diluted methylene blue were analysed in the visible ultraviolet, with two optical absorption peaks appearing at 611 nm and 663 nm. To study the catalytic effect of biosynthesised silver nanoparticles *Ag2O* NPs, 2.25 ml of NaBH4 solution (6.10^{-6} M) (considered as a reducing agent) was added to 3 ml of methylene blue solution (2.10^{-6} M), then biosynthesised silver nanoparticles (80 mg/l) were added. The degradation process was monitored by spectrophotometry in the wavelength range 400-800 nm every 5 min for 30 min. Decolourisation was observed by decreasing the absorbance of the solution at the maximum wavelength ($_{Amax}$). The experiments were used evaluate catalytic efficiency of biosynthesised *Ag2O* NPs.

III. RESULTS AND DISCUSSION Results and discussion

III.1 Characterisation silver oxide nanoparticles *NPs-Ag2O*

III.1.1. UV- vis spectroscopy and optical band gap

Preliminary studies suggest that phytochemical screening of the *H. hirsuta* plant extract reveals the presence of polyphenols, flavonoids and condensed tannins [147]. The as-prepared colloidal solution *of Ag2O* and the plant extract were analysedës by UV-Vis spectroscopy (see Figure Ш.1). In fact, the spectrum of the végëtal extract showed two peaks at 300 nm and 670 nm. After the reaction, the spectrum of *Ag2O* NPs has гёуёк a sitire peak at 430 nm. This peak corresponds to the characteristic absorption band of the *Ag2O* surface plasmon resonance [148].

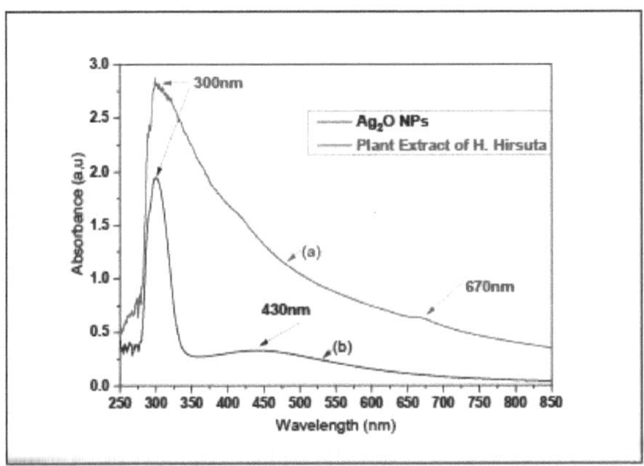

Figure III. 1: UV-visible spectrum (b) of biosynthëtisëes Ag2O NPs and
(a) of the vëgëtal extract of *H. Hirsuta.*

- Determination of the optical band gap

In дё^ тёга!, the optical band gap of a semiconductor can be dëterminatedëe by tracing the absorption coefficient as a function of photon energy, which can be estimëe using Tauc's formula Equation Ш.1)[149] :

$$(\alpha h v) = K(h v - E_g)^n \qquad (III.1)$$

Where a is the absorption coefficient, hv is the energy of the incident photon, K is a constant, Eg is the optical bandgap in ëlectronvolts (eV) and n is an exponent which can take two values depending on the nature of the ëlectronic transition, i.e.; n = 2 for a direct transition, and n = 1/2 for an indirect transition as shown in figure III. 2 and figure III.3 [150], [151].

- Urbach energy estimates

Urbach energy refers to the width of the band tails of the localised ëtats. The Urbach energy is determined from liquidation (Ш.2) of the slope of the boundary part of the Ияеё of ln(a) as a function of the energy of the *hv* photons (figure Ш.4) [152]. Table III.1 represents the values of the direct band gap, indirect band gap and Urbach energy of biosynthëtisëed *AgrO* NPs.

$$\ln(\alpha) = \frac{h v}{E u} + ln(\alpha_0) \qquad (III.2)$$

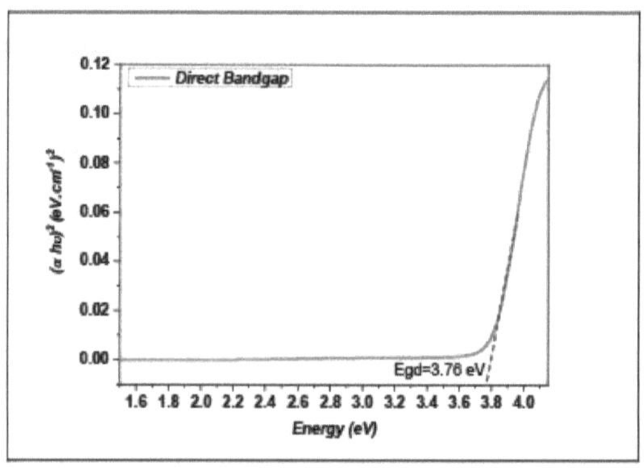

Figure III. 2: Dëtermination of the optical band gap for the direct a
using the Tauc method.

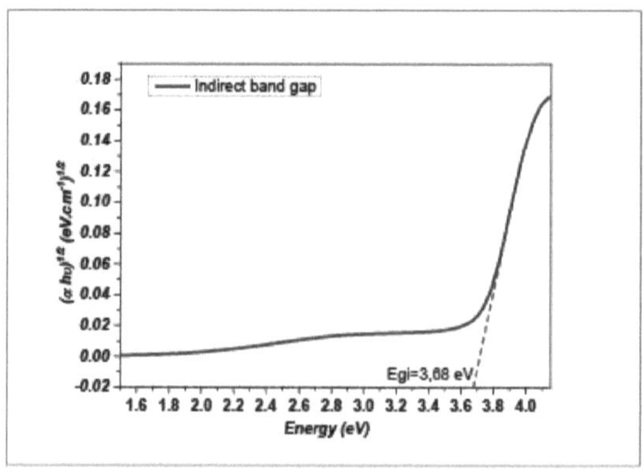

Figure III. 3: Dëtermination of the optical bandgap for the indirect a
using the Tauc method.

Table III. 1: Values of the direct and indirect band gap and the Urbach energy of
biosynthesised Ag2O NPs.

Energy (eV)		
Direct optical bandgap	Indirect optical bandgap	Energy from Urbach
3.76	**3.68**	**1.23**

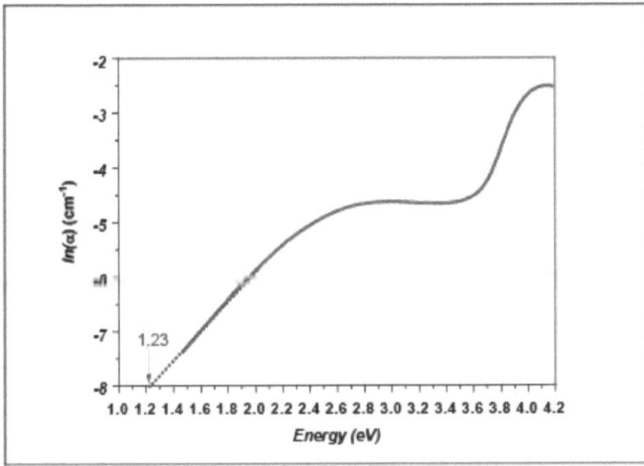

Figure III. 4: Tracë of ln(a) as a function of l^energy: Estimation of Urbach l^energy for biosynthëtisëes Ag2o nanoparticles.

Ш.1.2. Fourier transform infrared spectroscopy (FTIR)

Identification by FTIR analysis shows the potential presence of reducing and stabilising biomolecules in the extract of the H. Hirsuta plant extract. The FTIR spectrum obtained (Figure III.5) shows several absorption bands corresponding to the functional groups of the biomolecules present in the H. Hirsuta plant extract. Seven main absorption peaks were observed, the broad peak centred on 3400 and 3533 cm^{-1} is attributed to the O - H stretching vibrations of flavonoids and alkaloids, the intense peak at 1123 and 1687 cm^{-1} is due to the C = O stretching and N - H bending vibrations of the primary amide group which often exists in proteins [153]. The two peaks located at 1392 cm^{-1} and 1621 cm^{-1} are attributed to the NO and COC stretching vibrations of the aromatic ring respectively [154]. In addition, peaks at 673 cm^{-1} and 650 cm^{-1} are attributed to Ag-O and C-H out-of-plane bending vibrations respectively [155].

The results of the FTIR analysis show that the extract of the plant *H. Hirsuta plant extract* contains many different functional groups such as carboxyls, carbonyls, amides and phenols, which could be used as bioreducing agents and covering agents for the synthesis of *Ag.O* NPs [147].

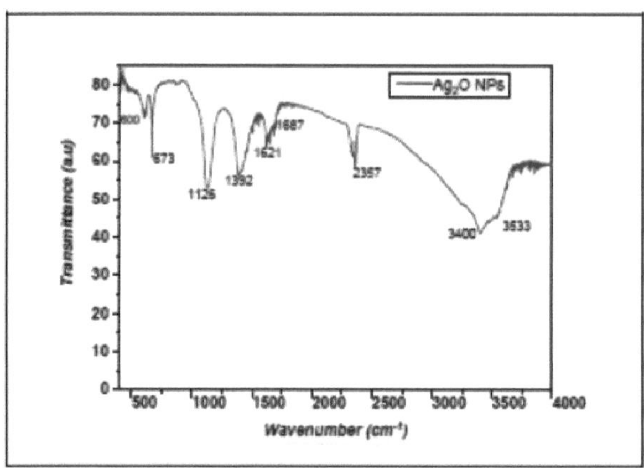

Figure III. 5 : FTIR spectrum of biosynthesised Ag2O NPs.

III.1.3. X-ray diffraction (XRD)

The X-ray diffraction pattern of biosynthesised Ag2O NPs is shown in Fig. 6. Several Bragg reflection peaks were observed in the XRD diffractogram located at values 2 of 27.28° , 33.36° , 55.93° , 66.52° and 70.01° corresponding to the (110), (111), (220), (311) and (222) planes of Ag2O silver oxide with a face-centred cubic crystal structure (JCPDS, File No. 00-012-0793) [138].

The crystalline size of the biosynthetic Ag2O NPs nanoparticles was calculated using the Scherrer formula (Equation (III.3)), considering the most intense peak located at the 0 value of 33.45°:

$$D = \frac{0.9 \times \lambda}{\beta \times cos\theta} \qquad (III.3)$$

Where D is the crystal size (nm), 0 is the total width at half the diffraction peak maximum (FWHM) of the strongest diffraction peak, X is the X-ray wavelength (for CuKa!= 1.5406 A), and 0 is the Bragg diffraction angle. Its value was 17.16 nm, and the average crystal size is 15.51 nm.

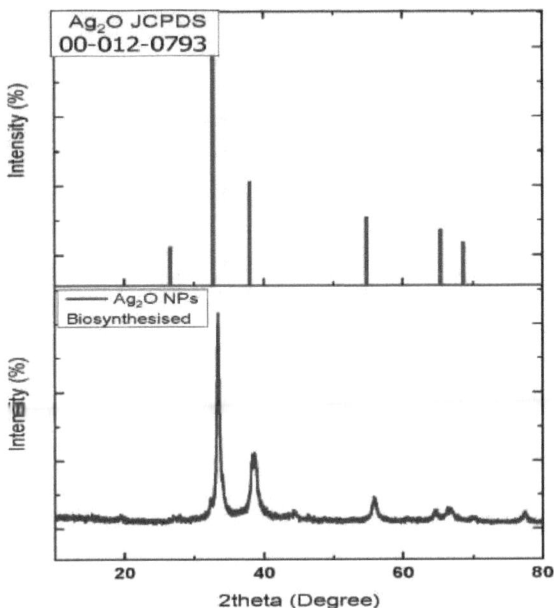

Figure III. 6: XRD diffractogram of biosynthëtisëes Ag2O NPs.

Ш.1.4 Morphological study by scanning electron microscopy (SEM)

SEM has ëlë usedë to ëstudy the morphology of Ag2O NPs and their morphological size. Figure III.7 (a-c) shows SEM images and ëlëmentary mapping of *Ag2O* NPs. It has ëlë observed that most of them have a sphërical shape. And in Fig. 7 (d), EDS analysis indicates that the ëlëments present in the synthëtisë product oxygëne O and silver Ag, i^Ument Ag ëbeing more abandon^ than i^Ument O.

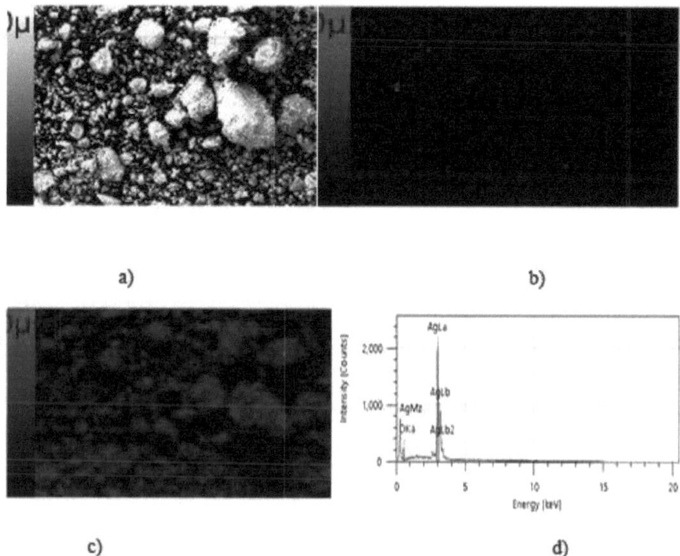

a)   b)

c)   d)

Figure III. 7: (a) SEM images of biosynthesised Ag2O spheres, (b,c) corresponding elemental
EDS mapping of O and Ag, (d) EDS spectrum of biosynthesised Ag2O NPs.

III.2 Catalytic activity for BM degradation

Catalytic hydrolysis of the MB dye in the presence of NaBH4 was examined to confirm the catalytic activity of Ag2O NPS. Catalytic degradation is monitored by UV-Vis spectroscopy. By adding the Ag2O nanoparticles to the reaction mixture, catalytic reduction of the dye takes place after a few minutes. The strong blue colour of the MB solution fades to colourless after 35 minutes during the degradation process.

The degradation efficiency is calculated by the following formula (Equation (III.4)):

$$D\% = \frac{(C_0 - C_t)}{C_0} \times 100 \qquad \text{(III.4)}$$

Where C_0 is the initial concentration of BM and C_t the immediate concentration in the sample.

The presence of amide groups of *Ag2O* silver oxide in the d^lectron transfer! from BH4^c) anions to m&hylene blue (BM) cations, which increases with time, which ë!т! ëalso similar to previously reported prëcëd *Ag2O* nanoparticles^ [144], [156]. The initial absorption peak at 663 nm has ë!ë attënuë with time, which dëmontre i'activile catalytique du produit des NPs d'Ag2O (figure III.8). Figure III.9 schematically shows the mëchanism of the catalytic dëgradation process of the BM dye. The number of cycles of use of biosynthëtisëed *dAg2O* nanoparticles for MB photocatalysis is ëgal to 4 (Figure III.11).

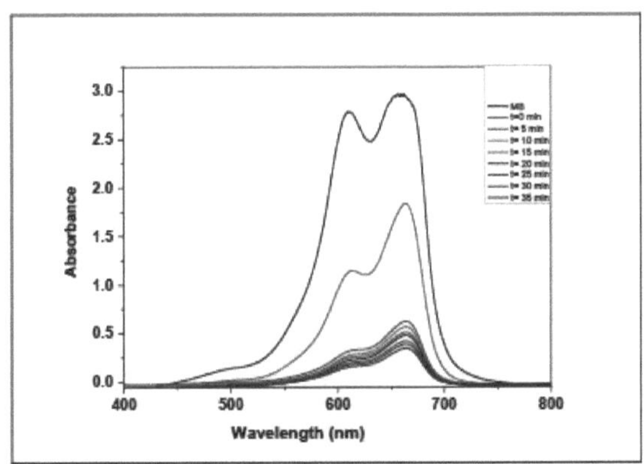

Figure III. 8: UV-vis spectrum for MB reduction using biosynthesised Ag2O biosynthëtisëes NPs.

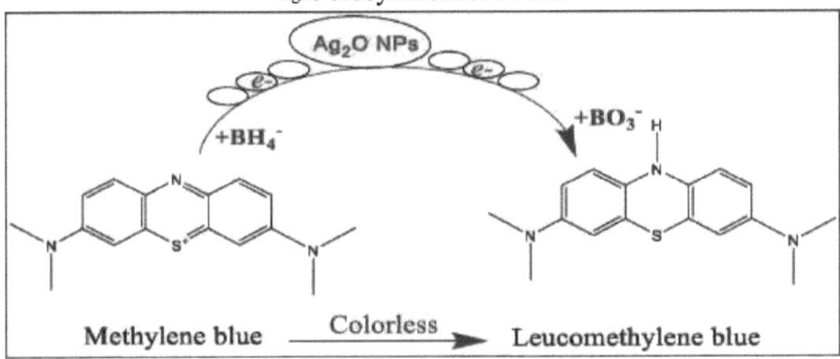

Figure III. 9: Scheëma of the catalytic dëgradation process of MB dye by Ag2O NPs.

Analysis of the kinetics of the dëgradation reaction has shown that the kinetics of the reaction is pseudo-first order. The reaction rate is determined by the following relationship: $ln(C_t/C_{(o)}) = ln(A_t/A_{(a)}) = -K_{app} \times t$, where $C_t(A_t)$ and $C_o(A_o)$ represent the concentration (absorbance) of the BM dye after and before degradation, respectively. The slope of the curve determines the value of Kapp (min^{-1}). The line plot of $ln(C_t/C_{0}\sim)$ versus time (Figure III.12) confirms the kinetic theory, where the Kapp value is 0.02 min^{-1} [144], [157]. The catalytic performance of silver oxide is linked to their specific morphology, which allows the rapid movement of electrons on the surface of silver oxide nanocatalyst, and their small size (the average size is 15.51 nm) ensures a large specific surface area that facilitates the dye degradation reaction by the nanocatalyst, which reaches the rate of 88.6% after a period of 35 min (Figure III.10). Thus, these two characteristics of the biosynthesised Ag2O nanoparticles accelerate the methylene blue dye degradation process.

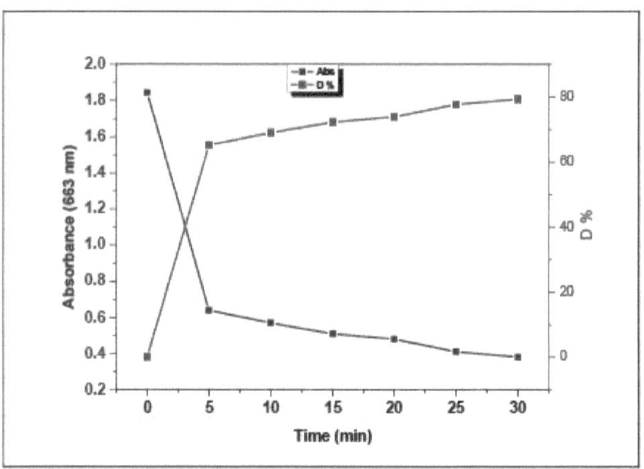

Figure III. 10: Evolution of the absorbance (663 nm) and degradation (D%) of the dye methylene blue dye as a function of time.

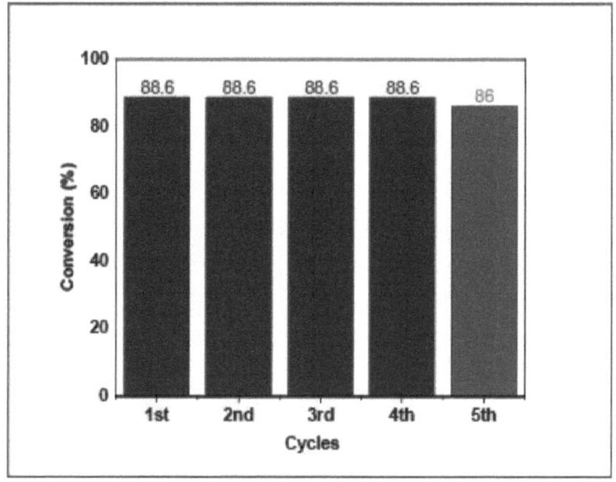

Figure III. 11: Percentage conversion of тё blue^ y1eпe in each catalytic cycle after 35 min.

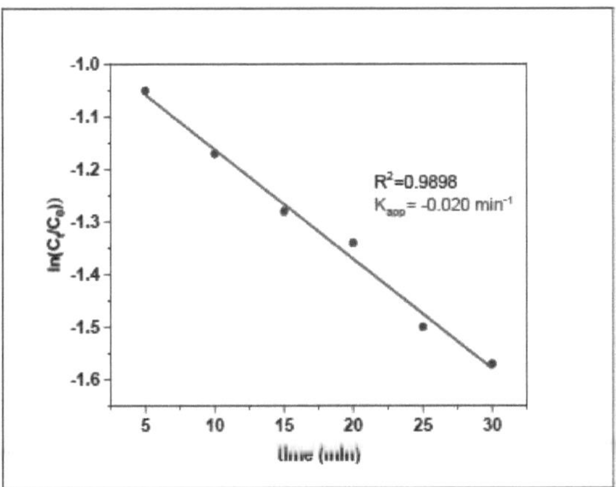

Figure III. 12: Traeë of ln(C_t/C_0) as a function of time for the catalytic reduction reaction of BM with Ag2O NPs.

IV. Conclusion

In this study, for the first time, the green synthesis of silver oxide was successfully achieved using the aqueous extract of the wild plant *H. hirsuta* . The process is easy, fast, inexpensive, environmentally friendly and requires no organic solvents or surfactants. As a result, this synthesis method is more advantageous than traditional methods for synthesising *Ag2O* nanoparticles. The shape of the biosynthesised *Ag2O* nanoparticles is almost spherical, cubic crystalline in nature, and the average crystal size is 15.51 nm, which implies that they will be of great interest to the research community. In addition, the *Ag2O* nanoparticles prepared have a better catalytic activity for the degradation of the BM dye under standard conditions, approaching 89% due to their morphology and small size.

Biosynthetised *Ag2O* nanoparticles are of great importance in wastewater treatment (degradation of dyes), medicine, cosmetics, paints, plastics and textiles.

CHAPTER IV

Biosynthesis and characterisation of nanoparticles of nickel II NiO and their high-efficiency photocatalytic application

I. Introduction

Today, many environmental problems have arisen as a result of the impact of various natural and man-made factors on the earth's crust. In general, any unnecessary and unacceptable modification of the environment due to various human activities is called pollution. The direct or indirect alteration of the biological, chemical and physical properties of natural water flows has harmful effects not only on human life, but also on aquatic ecosystems. There are different types of factor responsible for polluting natural water bodies, such as the population explosion, urbanisation, rapid industrialisation, water pollution, etc. Of all the factors involved rapid industrialisation, the textile, food, dyeing and paper industries are the main sources of water pollution. The effluents from these industries contain various types of dyes and organic pollutants. Contamination by organic dyes is thought to play an active role in the pollution of aquatic ecosystems. Approximately 10-15% of the world's total dye production is lost to wastewater during various processes [158]. Dyes present in wastewater prevent sunlight from penetrating watercourses, thereby reducing photosynthetic reactions. Some dyes are lethal, even neoplastic and malignant, and can pose a serious threat to human and animal health [159]. The contamination of industrial effluents by BM and Rh B dyes is a major problem and has a harmful effect on both types of ecosystem. With water scarcity and awareness of the threats posed by industrial effluents, international environmental standards are imposing many stringent requirements worldwide, leading to the development of innovative frameworks and techniques to remove dyes and other organic pollutants from wastewater before discharge [160]. In recent years, in this rapidly evolving technological era, nanotechnology has flourished, generating a wealth of scientific ideas that compete with the everyday challenges of rapidly developing technology [161]. Nanomaterials have attracted a wide range of scientific and technological interest because of their innumerable applications and specific properties [161]. These properties are conferred by their characteristic size, shape and surface type. Among these nanostructures, TiO2 is the most widely studied [162]. Recent studies on the photocatalytic activity of numerous p-type semiconducting transition metal oxides such as NiO, Cu2O, FeO, etc. have appeared in the literature [163], [164].

In recent years, several ëtudies have ëlë been conducted in the field of nanotechnology, using green plant materials and extracts for the biosynthesis metal oxide nanoparticles in order to avoid chemicals responsible for environmental pollution [165]. The biosynthesis of metal nanoparticles follows a green process using different plant fractions as reducing and stabilising agents. Among the many metal nanoparticles studied, nickel oxide is a valuable material used in various fields because of its unique properties [166]. It is therefore of great importance in the field of nanomaterials. Its properties include photoelectricity, catalysis and drug delivery, as well as the cathode in rechargeable batteries [167]. The increase in industrial activity always implies a great deal of pollution of the environment by chemical products, because due to the inadequacy of treatment systems, it is urgent to find simple and less expensive solutions, among which we find nanoparticles, which have already shown their potential application in the treatment of organic pollutants such as methylene blue and rhodamine B, widely used in several fields: chemistry, pharmacology, medicine, biology and textiles [168]. The use of these substances without the user's knowledge causes serious damage to human health and the environment [168]. Green synthesis is a modern field of biotechnology that represents an ecological and economical alternative to chemical and physical processes that are often harmful to the environment. In this method, natural reagents are biologically harmless, non-toxic and environmentally friendly [166], Cymbopogon citratus [169], Petiveria alliacea L. [170], Paeonia emodi [171], Ziziphora clinopodioides [172], Callistemon lanceotus (Myrtaceae) [173], [174], Centella Asiatica and Tridax [175], and many others have been used in the biosynthesis metal oxide nanoparticles [176].

For the green synthesis of metal oxide , the researchers used plant extracts that are widely available in nature. Because of their speed, environmental effectiveness and low cost, these plants are known as 'vital plants'. They have the power absorb minerals while respecting safety levels [177]. These methods include algae and microbes such as fungi, bacteria and viruses as reducing agents [174], [178]. The current method has more than one advantage: it is a cost-effective technique that does not use solvents or surfactants [179].

Studies on the use of nanostructured NiO as a photocatalyst for the degradation of organic dyes such as BM, Rh B, methyl orange and acid red 1 have ëlë reported [180], [181]. These reports indicate the potential applications of nanocrystalline NiO as a photocatalyst for many other factions, including the dëgradation of organic pollutants for water purification, as well as the reduction activity of 4-nitrophënol [182], [183]. NPs-NiO are chemically stable and exhibit very ëlevëous ëlectro-optic performance with a wide band gap (3.6-4.0 eV). NiO is a p-type semiconductor widely usedë in many chemical and physical applications such as catalysis, solar cells and gas dëtection [184], [185]. According to the literature, there are several mëthods to prepare NPs-NiO, such as thermal dëcomposition [186], combustion [187], sol-gel [188], [189], co-precipitation [190], spray pyrolysis [191], and the anodic arc plasma mëthod [192].

The present ëstudy focuses on simple chemical synthesis, which has the advantage over other mëthods of being simple, fast and ëenergy-efficient [193]. Extensive research around the world is attempting use industrial and natural precursors that are gënërally expensive. Despite this, the extract of *H. Hirsuta* has been used in this study due to its accessibility and price. The product obtained presents the best photocatalytic activ^ vis-a-vis MB and Rh-B dyes, which are more widely used in various industrial fields. We report the synthesis and cara^risation of single crystal NiO with a spëcific ëlevëe surface by a simple and direct green synthesis. The comparative study of the photocatalytic performances of -NiO NPs for the dëgradation of BM and Rh- B dyes has ëtë rëalisëed by adopting different optimised parameters, namely: the influence of pH, irradiation time and initial dye concentration. This is the first report of a ëtude comparative study on the photocatalytic performance of NPs-NiO biosynthëtisëesized by the vëgëtal extract of *H. Hirsuta*. In addition, the reducing propriëtë of 4-nitrophënol has ëlë ëtudiëe.

II. Materials and methods

II.1. . Equipment

The products used are : Nickel (II) nitrate $Ni(NO_3)_2,6H_2O$ (98%, MONTPLET& ESTEBAN SA BARCELONA.MADRID. Spain), the plant *H. Hirsuta* has been recoded in the province of Ouezzane in northern Morocco and stored in the dark, sodium hydroxide (98%, LOBA CHEMIE PVT.LTD. India). The glassware used was washed with acëtic acid and rinsed with distilled water. Two mortars, one made of porcelain and the other of zirconium, and a 1200 °C laboratory muffle furnace.

II.2. . Preparation of 1 extract of the plant *H. Hirsuta*

The plant *H. Hirsuta* was harvested in June, dried and stored in the dark at room temperature. After about two months, the *H. Hirsuta plant* was washed twice with distilled water and dried at room temperature for 48 hours, then ground to a fine powder. 10 g of *H. Hirsuta* was added to 400 ml in a 500 ml beaker. The mixture was shaken vigorously at 5,000 rpm and at room temperature overnight. The extract obtained was filtered through a piece of cloth and stored in a container. Finally, the filtrate was centrifuged at a speed of 10,000 rpm at room temperature to obtain a light brown supernatant and stored in a light-tight container for later use.

II.3. . Biosynthesis nickel oxide nanoparticles (NPs-NiO)

NPs-NiO were synthesised by the coprecipitation method, using extract from the plant *H. Hirsuta plant extract*. In a 250 ml beaker, 20 ml of $Ni(NO_3)_2, 6H_2O$ (5 mM) nickel (II) nitrate solution was added to 60 ml of *H. Hirsuta* plant extract with an extract/salt ratio of (3:1). *Hirsuta plant* extract with an extract/salt ratio (3:1) under stirring at 7000 rpm at room temperature for 10 min. The formation of biosynthesised NPs-NiO was monitored by the change in colour and the appearance of a green precipitate after 5 min and confirmed by UV-Vis spectroscopy after adjusting the pH = 8 with a sodium hydroxide solution (0.1 M). Finally, the mixture was centrifuged at a speed of 10,000 rpm at room temperature to remove the supernatant, the precipitate was washed twice with distilled water and methanol, dried in an oven at a temperature of 70°C for 24 h, and calcined at 500°C in a muffle furnace being placed in a glass container and stored for further analysis and applications.

II.4. . Characterisation of biosynthesised nickel oxide (NPs-NiO)

II.4.1.. UV-visible spectroscopy

The production of biosynthesised NPs-NiO was monitored using UV-visible spectra (DR 6,000 spectrophotomëtre with RFID technology (HACHLANGE, GERMANY)). Measurements were carried out under normal temperature and pressure conditions (20°C and 1atm) in the wavelength range 250-800 nm using a quartz cell.

II.4.2.. X-ray diffraction (XRD)

The crystalline structure of biosynthëtisëes NPs-NiO has ël.ë dëterminëe using a powder X-ray diffractometer (PhaserD2 diffractometer, Broker, USA) with $Cu-K_a$ radiation of wavelength X = 0.15406 nm in the range 10°-80°, operating at 30 kV and 10 mA.

II.4.3.. Fourier transform infrared spectroscopy FT-IR

Various functional groups were observed by FTIR spectral analysis (Varian 800 / Gladiatr modële (Scimitar series, Australia / Pike Technologies, USA) such as carbonyls, polyphënols and amides, which could be reducers for the biosynthesis of NiO-NPs.

II.4.4.. Scanning electron microscopy (SEM/EDS)

The morphology and shape of biosynthëtisëed NPs-NiO were ëtë ëtudiëes by scanning electron microscopy coupled to EDS (SEM-JEOL IT500HR) with the following propriëtës: landing voltage 10.0 kV, WD 11.0 mm, mëquantification method ZAF, magnification x8500, high vacuum modeë.

II.5. . Catalytic activity of biosynthesised NPs-NiO

II.5.1.. Catalytic degradation of methylene blue and rhodamine B dyes

In a typical expërience, 1 mg of synthëtisëes NPs-NiO was ëtë addedë to 50 ml (5 mg.L^{-1}) of an aqueous solution of BM and Rh B. Then, the solution was ë1ë maintained in a Mayer earlen (reactor) with continuous stirring to ensure that the catalyst suspensions ëtended uniformly throughout the reaction time. Meanwhile, after adjustingë the pH (pII=10 fɯ MD, ɑɯd pII=2 fuɪ Rh B), un ɑdcquɑtc quɑnthc uɪ dye suluɪuɪ was ëɪë withdrawn and filtered to sëparate the NPs-NiO. Next, the filtered solution was analysed using a UV-Vis spectrophotometer at the maximum absorption wavelength of 663 nm and 554 nm of the dye to obtain the concentration of BM and Rh-B in the solution. The percentage of dye degradation was calculated using liquidation (IV.1).

$$D(\%) = \frac{A_0 - A_t}{A_0} \times 100 \qquad (IV.1)$$

Where A0 and At represent the absorbance of the radiation and at time t, respectively.

II.5.2.. Reduction of 4-nitrophenol (4-NP) by NPs-NiO

The 4-nitrophenol (4-NP) reduction experiments were carried out as : An aqueous solution of 4-NP (60 ml, 5 10^{-5} M) was mixed with a freshly prepared NaBH4 solution (1.5 10^{-4} M, 15 ml), giving a dark yellow solution. Next, 5 mg of catalyst (prepared as above) was dispersed in the solution, followed by sonication for 1 minute. After adjusting pH =9, the progression of the reaction for all experiments was monitored using a UV-vis spectrophotomëtre [182]. The reuse of NPs-NiO was carried out for 3 cycles, at the end of the firstëreduction experiment, the catalyst was collected by centrifugation, washed with water and ethanol, then oven dried at 80°C for 3 h for the next cycle.

III.Results and discussion

III.1. 1. Characterisation nickel oxide nanoparticles (NPs-NiO)

III.1.1. 1. UV- vis spectroscopy and optical band gap

Preliminary studies suggest that phytochemical screening of *H. Hirsuta* plant extract detects the presence of polyphenols, flavonoids and condensed tannins [194], [195]. The colloidal solution of Ni(OH)2 was therefore prepared using the plant extract and analysed by UV-Vis spectroscopy (Figure IV.1(a-b)). The spectrum of the plant extract showed two peaks at 300 nm and 670 nm. After the reaction, the spectrum obtained shows that there is a peak at 390 nm which corresponds to absorption band characteristic of the surface plasmon resonance of the Ni(OH)2 precipitate in solution [196]. Figure IV.2 shows the UV-vis spectromëtre of NPs-NiO obtained after calcination, the peak indicated at 301 nm is attributed to NPs-NiO [197].

- *Determination of the optical band gap*

In general, the optical bandgap of a semiconductor can be determined by plotting the absorption coefficient as a function of photon energy, which can be estimated using the Tauc formula (equation (IV.2)) [198]:

$$(\alpha h v) = K(h v - E_g)^n \qquad (IV.2)$$

Where a is the absorption coefficient, hv is the energy of the incident photon, K is a constant, Eg is the optical bandgap in electronvolts (eV) and n is an exponent which can take two values depending on the nature of the electronic transition, i.e. n = 2 for a direct transition and n = 1/2 for an indirect transition, as shown in figure IV.3(a-b) [199], [200].

*   ***Urbach energy estimates***
The Urbach energy is Hëe to the width of the band tails of the localised ëtats. The Urbach energy is dëterminatedëe from liquation (IV.3) of the slope of the boundary part of the Изсё of ln(a) as a function of the energy of the hv photons (Figure IV.4) [201]. The results are presented in Table IV.1, and a report on the bandgap energy values of NPs-NiO in Table IV.2.

$$\ln(\alpha) = \frac{h v}{Eu} + \ln(\alpha_0) \qquad (IV.3)$$

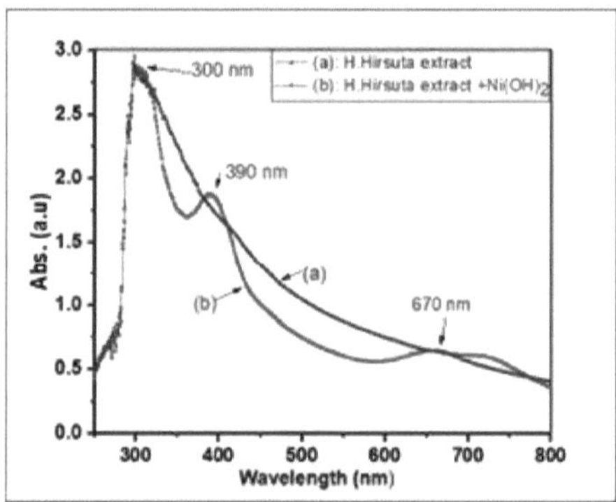

Figure IV. 1: UV-visible spectrum of (a): the vëgëtal extract of *H. Hirsuta,* and (b): of Ni(OH)2 in solution.

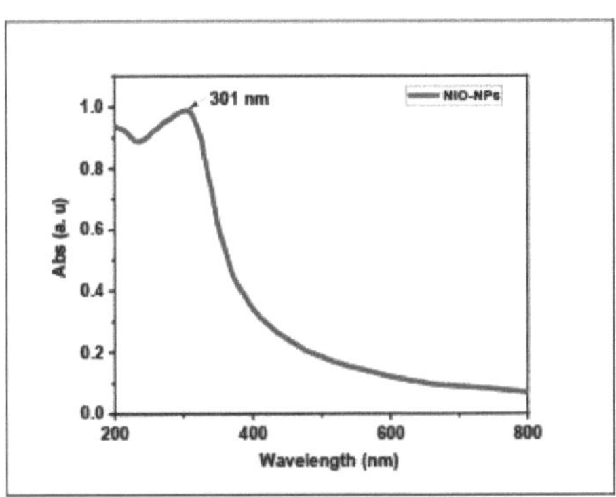

Figure IV. 2: UV-visible spectrum of calcined NiO NPs at 300°C.

Table IV. 1: Values of the direct and indirect band gap and of Urbach energy of biosynthëtisëes NPs-NiO.

Energy (eV)		
Direct band	Indirect bandwidth	Energy from Urbach
3.02	3.40	2.00

Table IV. 2: Band gap energy ratios for NiO NPs.

Sample No	Gap energy	Refs.
1	3.30	[202]
2	3.41	[203]
3	3.51	[204]
4	2.51	[205]
5	3.40	[206]
6	3.47	[205]
7	3.13	[207]
8	3.00	[208]
9	3.47	[209]
10	3.40	**Present work**

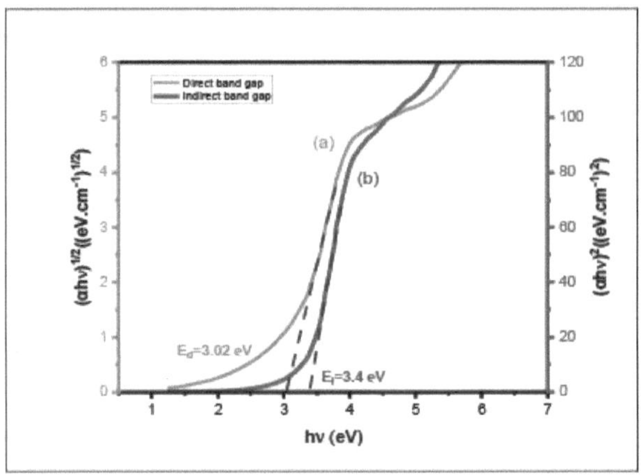

Figure IV. 3: Dëtermination of the optical band gap for (a) the direct transition and (b) the indirect transition using the Tauc mëthod.

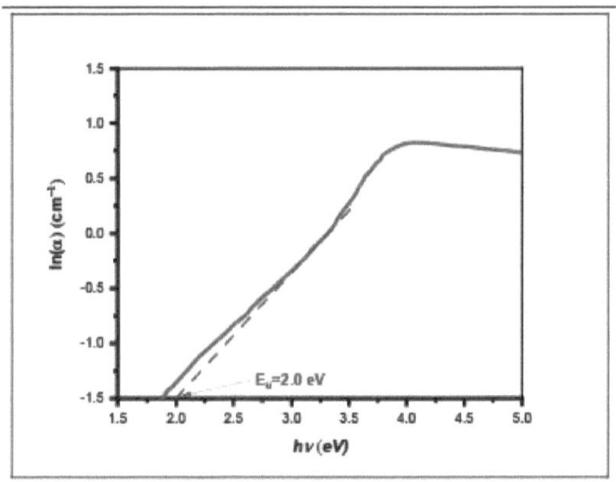

Figure IV. 4: Tracë of ln(a) as a function of l^energy: Estimation of the Urbach energy. for NPs-NiO biosynthëtisëes.

Ш.1.2. Fourier transform infrared spectroscopy (FTIR)

Identification by FTIR analysis shows the potential presence of reducing and stabilising biomolecules in the extract of the *H. Hirsuta plant extract*. The surface functional group of the synthesised NiO NPs was obtained by FT-IR transmission spectra and is shown in Figure IV.5.

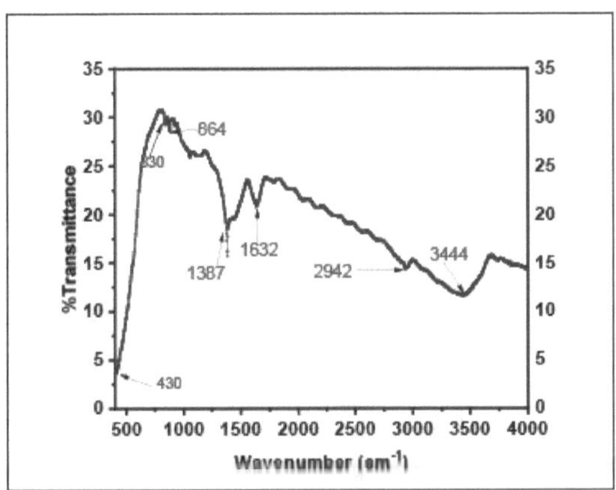

Figure IV. 5: FTIR spectrum of biosynthesised NPs-NiO nanoparticles.

In дёпёral, the absorption band interatomic vibrations, in particular of mёtalic oxides, has bands below 800 cm-1, the peak recorded at 432 cm-1 represents the appearance of Ni-O bonds [199], [203], while the double band at 1387 and 1632 cm-1 corresponds to the vibration mode of aromatic and carbonyl groups, respectively. The peak observed at 2942 cm-1 is Kё to the vibration of C-H bonds [210], and broad absorption band with a centre peak at 3444 cm-1 is attributed to the vibration of the hydroxyl group [182]. Thus, the spectrum obtained shows that the NPs-NiO synthёtisёed in this manner are of a very pure phase and confirms the result of the XRD analysis.

III.1.3. X-ray diffraction (XRD)

X-ray diffraction is a viable tool for characterising the solid-state structure of nanomaterials. DRX analysis has been used to characterise nickel oxide after calcination at 500°C. The XRD diffractogram of the NPs prepared by the green route is shown in Figure IV.6. The diffraction peaks observed at angles of 37.25, 43.41, 62.95, 75.43 and 79.25 are related to the (111), (200), (220), (311) and (222) crystal planes. Comparing the XRD diffractogram of the synthёtisёed nanoparticles with the standard sample, the final diffraction diffractogram (JCPDS card no. 01-073-1523) [211], revealed that the biosynthёtisёed NiO NPs have a cubic crystal structure and space group of (Fm- 3m) [212]. The size of the biosynthёtisёes NPs-NiO nanoparticles has ёlё calculated using the Scherrer formula Eq. (IV.4) consideringёranting the most intense peak at value 2θ of 43.414°

$$D = \frac{0.9 \times \lambda}{\beta \times cos\theta}$$

(IV.4)

Where D is the crystallite size (nm), θ is the total width at the maximum diffraction peak mask (FWHM) of the most intense diffraction peak, θ is the Bragg diffraction angle, and λ is the X-ray wavelength (for Cu-K$_a$, λ= 1.5406 A). Its value ёtait 24.3 nm, and the average crystal size is 20.82 nm, and the structural paramёters are shown in Table IV.3.

Table IV. 3: Structural parameters of biosynthёtisёed NPs-NiO.

Pic no.	2θ (Degree)	Plans (hkl)	FWHM (P)	Crystallite size (nm)
1	37,252	111	0.12	17.8

2	43,414	200	0.096	24.3
3	62,952	220	0.264	16.4
4	75,431	311	0.12	23.7
5	79,257	222	0.336	21.9
Medium crystallite size				**20.82**

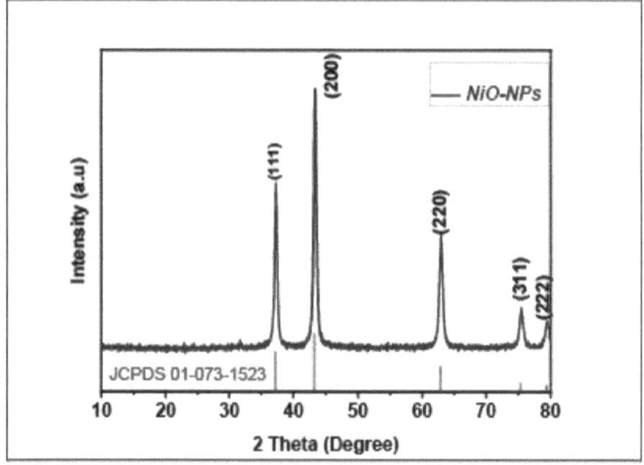

Figure IV. 6: XRD diffractogram of biosynthesised NiO NPs.

III.1.4 Morphological study by scanning electron microscopy (SEM/EDS)

SEM has ёlё usedё to ёstudy the morphology of NiO NPs and their morphological size. Figure 7(d) shows the EDS of NiO NPs. It is clearly indicated that the synthётisёed nanoparticles consist only of Ni and O. This confirms the purity of the NPs-NiO and other impurities have ёlё detected in the spectrum of the sample. The weight percentage observed from the energy dispersive spectra is presented in Table IV.4, which shows that the nanoparticles formed are rich in nickel relative to oxygёne, and confirms the purity of the biosynthesised NPs-NiO. SEM mapping images of the NPs-NiO are shown in Figure 7(b,c). These images show an irregular spherical morphology with different particle sizes due to agglomeration (Figure 7(a)). The surface area is high, which is beneficial for photocatalytic activity [185], [213].

Table IV. 4: EDS analysis of biosynthesised NiO-NPs.

Element	Mass (%)
Ni	95.57
O	4.43
Total	100

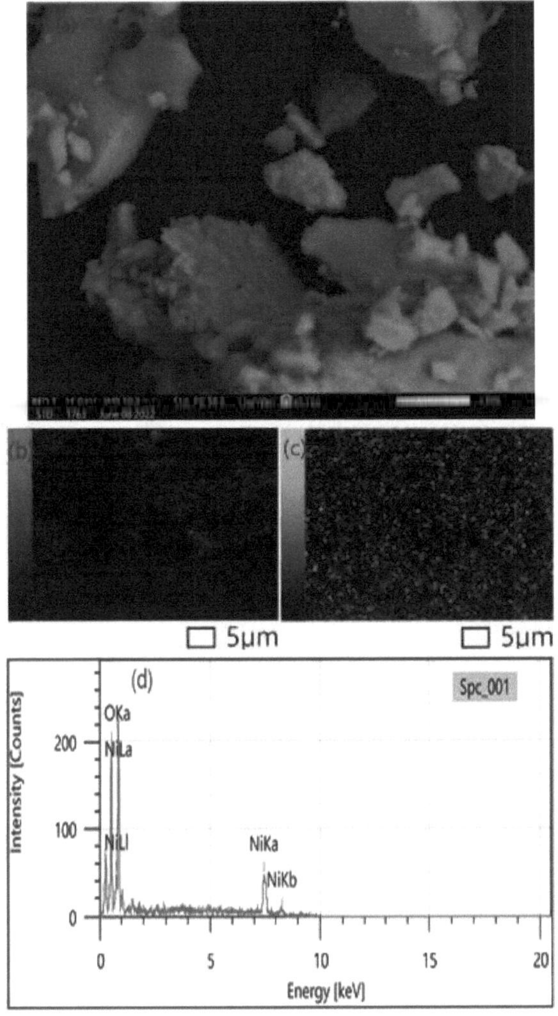

Figure IV. 7: (a) SEM images of the bюsynthëtisёes NiO-NPs, (b,c) Mapping images of the corresponding
ёlётепЫге corresponding EDS of O and Ni respectively, (d) EDS spectrum of NiO-NPs biosynthëtisёes.

III.2 Catalytic activity for the degradation of BM and Rh B

The comparative study of the efficiency of the NPs-NiO catalyst for the dёgradation of the BM and Rh B dyes was carried out under the conditions optimised by S. D. KHAIRNAR et al. (for BM at pH=10 and for Rh B at pH=2) with a dye concentration of 5 mg/L and a catalyst dose of 1 mg/L [158]. The results are shown in Figure IV.8(a-b). The degradation of BM showed a constant reduction with increasing irradiation time under visible light. Decolourisation of the dye solution occurred within 120 minutes of irradiation. The corresponding BM degradation rate was 97.19%. For Rh-B, degradation increased progressively with irradiation time and reached equilibrium. This is due to the formation of zwitterions that occupy the active sites of the catalyst. Consequently, the degradation of Rh-B does not exceed 79.42% at the same time (Figure IV.9(a-b)). The data were used for the cinetic study of the BM and Rh B dyes in the presence of NPs-NiO, which shows that the degradation reactions of the BM and Rh B dyes are pseudo-first-order reactions. Analysis of the kinetics of the degradation reaction showed that the kinetics of the reaction is pseudo-first order. The reaction rate is determined by the following relationship: $ln(C_0/C_t) = K \times t$, where c_t and c_o represent the dye concentration after and before degradation, respectively. The slope of the curve determines the value of K (min^{-1}) (Figure IV.10(a-b)). The photocatalytic activity can be compared with the k value and the linear regression coefficient (R^2) for the BM and Rh B solutions. The k values, which are obtained by linear fitting of each curve, are 2.073×10^{-2} min^{-1} for BM and 1.287×10^{-2} min^{-1} for Rh B. The photocatalytic degradation mechanism can be explained by the following equations.

$$NiO + h\nu \longrightarrow e^- + h^+$$

$$OH^- + h^+ \longrightarrow {}^{\bullet}OH$$

$$H_2O + h^+ \longrightarrow OH + H+$$

$$O_2 + e^- \longrightarrow {}^{\bullet}O^-$$

$$O_2^- \longrightarrow HO_2$$

$$2H_2O \longrightarrow O_2 + H_2O_2$$

$$H_2O_2 + O_2 \longrightarrow {}^{\bullet}OH + OH^- + O_2$$

$$Dye + {}^{\bullet}OH + {}^{\bullet}O_2 \longrightarrow CO_2 + H_2O + Inorg.salts.$$

The catalytic performance of NPs-NiO is Hëe to their irregular morphology that allows the rapid dëplacement of electrons on the catalyst surface, and their small size (average size of 20.82 nm) ensures a large spëcific surface area that facilitates the dye dëgradation faction by nickel oxide nanocatalyst, which reaches the rate of 97.19% and 79.42% after an irradiation of 120 min. These two characteristics of biosynthëtisëes NPs-NiO therefore accelerate the dëgradation process of BM and RhB dyes. The photocatalytic dëgradation efficiencies of BM and Rh-B dyes by some mëtalic oxide have been compared in Table IV.5.

Table IV. 5: Comparison of the photocatalytic dëgradation efficiency of the dyes BM and Rh B dyes.

Photocatalysts	Degradation BM	Rh B	Refs.
Fe2O3-CuO-ZnO	79		[214]
CdO-NiO-ZnO	86		[215]
CdO-ZnO		97.6	[216]
GO-Fe3O4-ZrO2		98	[205]

NiO-CdO-ZnO	98	99	[217]
NiO	60		[218]
NiO	**97.19**	**79.42**	**Present work**

Figure IV.11 describes a simple mëchanism that explains the enhanced photocatalytic activity of NiO NPs. The shape and effective surface area of nanoparticles are two crucial parameters that can amëiorate photocatalytic performance. A large surface area photocatalytic efficiency and reagent absorption. Collisions between sunlight and photocatalyst nanoparticles require less energy to excite electrons from the valence band (BV) to the conduction band (BC) in semiconductor space, producing more photons that stimulate valence band electrons [218]. Subsequently, this electronic excitation generates an equal number of holes in the valence band, and the stimulated electrons can directly or indirectly produce hydroxide radicals. Through the process of electron loss, OH hydroxide ions are converted into hydroxyl radicals (OH⁻), which are an integral part of the conversion of organic matter into minerals. This conversion is essential for the elimination of the dyes BM and Rh B. In addition, these organic compounds are decomposed by the loss electrons, leading to their transformation into H_2O and CO_2, which are ultimately released back into the atmosphere [202].

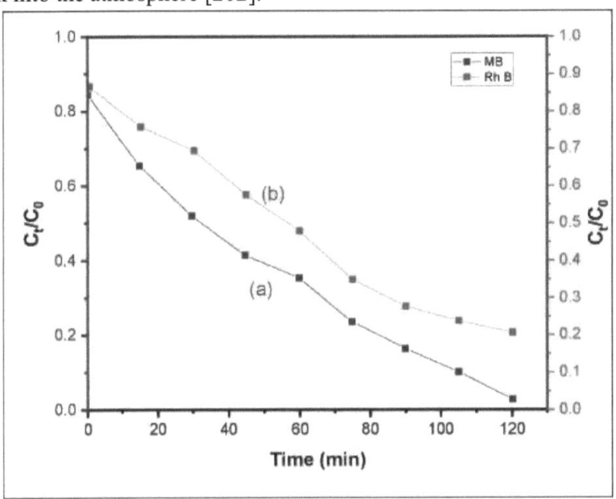

Figure IV. 8: Tracë of (C_t/C_0) as a function of time for the reaction of (a) BM, and (b) Rh B with NPs-NiO.

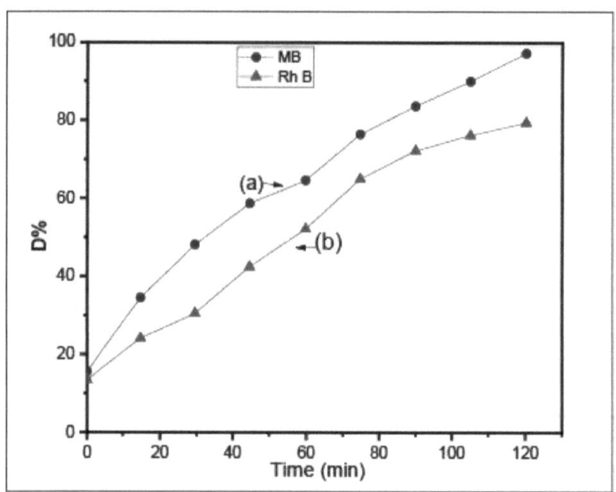

Figure IV. 9: Photocatalytic Dëgradation (a) BM, conditions pH=10, dye concentration dye 5 mg/L and catalyst dose 1 mg/L. (b) Rh B, conditions: pH=2, dye concentration concentration 5 mg/L and catalyst dose 1 mg/L.

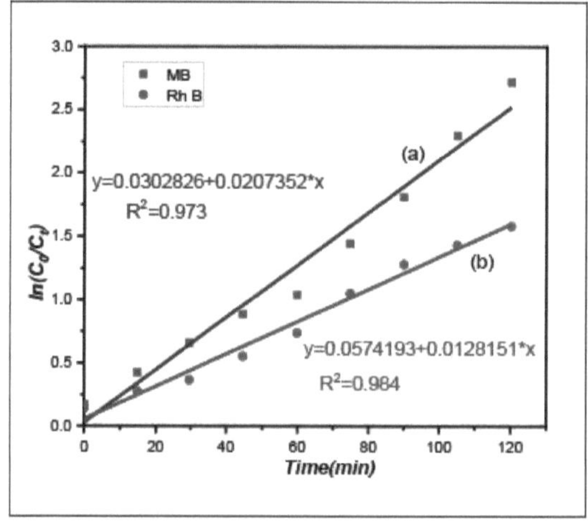

Figure IV. 10: Tracë of $ln(C_o/C_t)$ as a function of time for the catalytic reduction reaction of
catalytic reduction of (a) BM, and (b) Rh B with NPs-NiO.

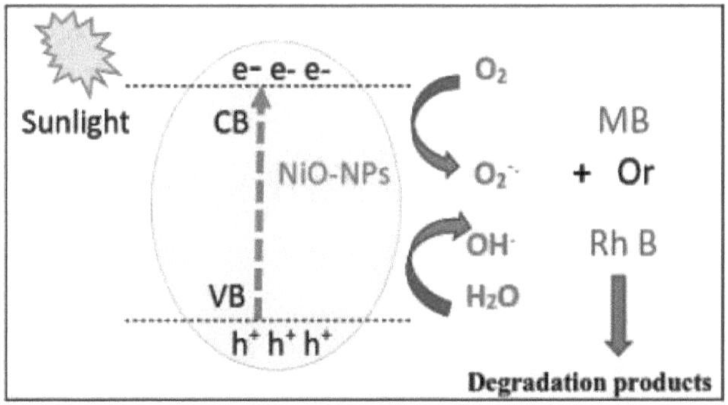

Figure IV. 11: Scheëma of the photocatalytic mëcanism of NPs-NiO.

III.3 Cutalytic reduction of 4-NP by NPs-NiO

The catalytic reduction of 4-NP to 4-aminophënol (4-AP) was ëlë chosen to ëstudy the catalytic activity of the prepared NiO NPs. The absorption band of 4-NP appears at 317 nm, when the freshly prepared $NaBH_4$ solution is addedëe, the light yellow colour of 4-NP changes to light yellow due to the formation of 4-nitrophenolate ions, and absorption band shifts to 400 nm. When reduction of 4-nitrophënolate begins, the intensity of the band at 400 nm decreases with the appearance a new band at 300 nm due to 4-AP. Furthermore, 4-NP is not reduced by $NaBH_4$ in the absence of catalyst. As can be seen from the UV-vis spectra (Figure IV. 12), the NPs-NiO show a reduction activity of 91.7% of 4-NP over 10 min. As an excessive amount of *$NaBH_4$* was used in these experiments, the reaction is therefore independent of the sodium borohydride concentration. The kinetic data are fitted to first-order NiO rate equation, and straight lines are obtained by plotting $\ln(c_t/c_0)$ against time (Figure IV.13) and their slope gives the value of the rate constant, which is 2.43 10^{-2} min^{-1}. The possible mechanism for the catalytic reduction of 4-NP using $NaBH_4$ on NiO-NPs occurs in four steps [219]: In the firstëre step, BH_4^- releases hydride ions into the aqueous medium, which bind to the NiO surface. In the second step, hydrogen is covalently bonded to the NiO surface. The rate-limiting step results in the adsorption of nitro groups onto the NiO surface (step 3). In addition, the adsorbed 4-NP and bound hydrogen atoms interact strongly. The hydride ion attacks the adsorbed nitro groups, electron transfer occurs from the BH_4^- donor to the 4-NP acceptor, followed by desorption of 4-AP into the aqueous medium.

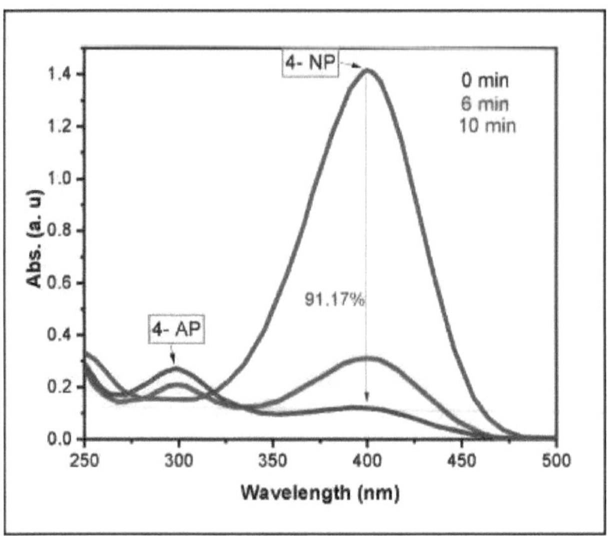

Figure IV. 12: UV-vis spectrum for the conversion of 4-NP to 4-AP using NPs-NiO.

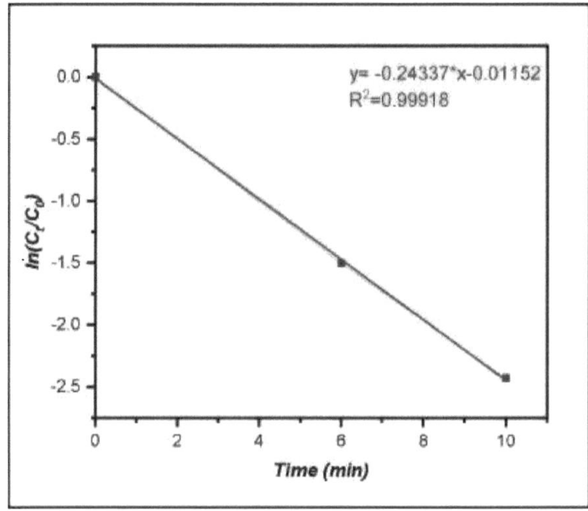

Figure IV. 13: Tracë of In (C_t/C_0) as a function of time for the reduction reaction of 4-NP with NPs-NiO.

IV. CONCLUSION

In this ëstudy, we successfully synthëtisë nickel oxide (NPs-NiO) using a simple and environmentally friendly green synthesis mëthod. The nanoparticles were ë1ë prepared using an extract of the plant *H. Hirsuta plant*, resulting in NPs-NiO with excellent stability and recyclability for Elimination of hazardous dyes. The biosynthëtisëed NPs-NiO have ë1ë caractërisëes using various analytical techniques, including UV-vis, XRD,

SEM, EDS and FTIR. Cara^risation has rëyë^   that the biosynthëtisëed NPs-NiO presented an irregular shape, with an average crystallite size of 20.82 nm. In addition, the nanoparticles possëd direct and indirect gap ënergies of 3.02 and 3.42 eV, respectively, with an Urbach ëenergy measured at 2.0 eV. These results provide valuable insights into the structural and optical propriëtës of biosynthëtisëed NPs-NiO.

CONCLUSION

Due to the significant environmental pollution affecting the world, green technologies and chemistry are becoming increasingly widespread and usedëes. The field of nanotechnology is advancing rapidly, employing sophisticated technologies and involving the construction and use of nanomaterials between [1 and 100] nm in size to meet the needs of industry. This is a unifying field where physics, chemistry and biology converge and intertwine. This field presents challenges in terms of structure, function and implementation, in which chemistry plays a particularly crucial role.

The application of nanotechnologies in biological fields is called nanobiotechnology. Chemists, physicists and biologists all see nanotechnology as an extension of their respective disciplines, and collaborative efforts in which each contributes equally are commonplace. The result is the hybrid field of nanobiotechnology, which makes extensive use of metal nanoparticles in a wide range applications due to their exceptional properties. These properties are mainly influenced by their size and shape.

In this ëtude, the green biosynthesis silver oxide $Ag2O$ and nickel oxide NiO nanoparticles using *H. hirsuta* plant extract were studied. The characteristics of these nanoparticles obtained were analysed using standard techniques such as UV-Vis, FT-IR, XRD, SEM and EDAX and an evaluation of their photocatalytic activities of the organic dyes BM and Rh B, as well as the reduction of 4-nitrophenol. Silver oxide $Ag2O$ and nickel oxide NiO were synthesised by the reduction reaction of silver and nickel ions due to the presence of phënolic compounds in *H. hirsuta* plant extracts that act as a bioreducing agent. The characterisation techniques used validated the production silver oxide $Ag2O$ and nickel oxide NiO nanoparticles.

UV-Vis and FTIR optical characterisation techniques confirmed the formation of $Ag2O$ silver and NiO nickel oxide discerning $Ag2O$ silver and NiO nickel oxide absorption bands, and the disappearance of polyphënol absorption bands induced by the reduction of métaℓℓque ions.

XRD analysis revealed the presence of a single phase of silver Ag2O and nickel oxide NiO in the biosynthëtisë nanoparticles obtained from the extract of the plant *H. hirsuta*: a single phase of cubic silver oxide Ag2O with a central face and an average grain size of 51.51 nm, and a single phase of nickel oxide NiO with a face-centred cubic structure, the crystallite size of the particles varying from 16.4 to 24.3 nm, with an average crystallite size of 20.82 nm. The nanoparticles synthëtisëed from *H. hirsuta* extract presented a single phase that coincides perfectly with the data that already exist in the bibliography.

Indeed, SEM images clearly show that Ag2O silver oxide agglomërate in an almost spherical form of variable sizes, but for NiO nickel oxide show an irreguliëre spherical morphology with different particle sizes due to agglomeration.

BIBLIOGRAPHICAL REFERENCES

[1] B. Mekuye and B. Abera, "Nanomaterials: An overview of synthesis, classification, characterization, and applications," *Nano Select,* vol. 4, no. 8, pp. 486-501, 2023, doi: 10.1002/nano.202300038.

[2] Y. Khan *et al*, "Classification, Synthetic, and Characterization Approaches to Nanoparticles, and Their Applications in Various Fields of Nanotechnology: A Review," *Catalysts*, vol. 12, no. 11, Art. no. 11, Nov. 2022, doi: 10.3390/catal12111386.

[3] S. Akin and S. Sonmezoglu, "Chapter 2 - Metal Oxide Nanoparticles as Electron Transport Layer for Highly Efficient Dye-Sensitized Solar Cells," in *Emerging Materials for Energy Conversion and Storage*, K. Y. Cheong, G. Impellizzeri, and M. A. Fraga, Eds., Elsevier, 2018, pp. 39-79. doi: 10.1016/B978-0-12-813794-9.00002-8.

[4] Y. Yoon, P. L. Truong, D. Lee, and S. H. Ko, "Metal-Oxide Nanomaterials Synthesis and Applications in Flexible and Wearable Sensors," *ACS Nanosci. Au*, vol. 2, no. 2, pp. 64-92, Apr. 2022, doi: 10.1021/acsnanoscienceau.1c00029.

[5] R. Khursheed *et al*, "Biomedical applications of metallic nanoparticles in cancer: Current status and future perspectives," *Biomedicine & Pharmacotherapy*, vol. 150, p. 112951, Jun. 2022, doi: 10.1016/j.biopha.2022.112951.

[6] S. Sim and N. K. Wong, "Nanotechnology and its use in imaging and drug delivery (Review)," *Biomedical Reports*, vol. 14, no. 5, pp. 1-9, May 2021, doi: 10.3892/br.2021.1418.

[7] M. P. Nikolova and M. S. Chavali, "Metal Oxide Nanoparticles as Biomedical Materials," *Biomimetics (Basel)*, vol. 5, no. 2, p. 27, Jun. 2020, doi: 10.3390/biomimetics5020027.

[8] F. D. Guerra, M. F. Attia, D. C. Whitehead, and F. Alexis, "Nanotechnology for Environmental Remediation: Materials and Applications," *Molecules*, vol. 23, no. 7, p. 1760, Jul. 2018, doi: 10.3390/molecules23071760.

[9] "Applications of Metal/Metal Oxides Nanoparticles in Organic Transformations," in *Materials Research Foundations*, 1st ed, vol. 83, Materials Research Forum LLC, 2020, pp. 134-156. doi: 10.21741/9781644900970-6.

[10] S. D. Anderson, V. V. Gwenin, and C. D. Gwenin, "Magnetic Functionalized Nanoparticles for Biomedical, Drug Delivery and Imaging Applications," *Nanoscale Res Lett*, vol. 14, no. 1, p. 188, May 2019, doi: 10.1186/s11671-019-3019-6.

[11] J. Singh, T. Dutta, K.-H. Kim, M. Rawat, P. Samddar, and P. Kumar, "'Green' synthesis of metals and their oxide nanoparticles: applications for environmental remediation," *Journal of Nanobiotechnology*, vol. 16, no. 1, p. 84, Oct. 2018, doi: 10.1186/s12951-018-0408-4.

[12] Z. Ullah *et al*, "Biogenic Synthesis of Multifunctional Silver Oxide Nanoparticles (Ag2ONPs) Using Parieteria alsinaefolia Delile Aqueous Extract and Assessment of Their Diverse Biological Applications," *Microorganisms*, vol. 11, no. 4, Art. no. 4, Apr. 2023, doi: 10.3390/microorganisms11041069.

[13] M. S. Chavali and M. P. Nikolova, "Metal oxide nanoparticles and their applications in nanotechnology," *SN Appl. Sci.* vol. 1, no. 6, p. 607, May 2019, doi: 10.1007/s42452-019-0592- 3.

[14] S. H. Lee and B.-H. Jun, "Silver Nanoparticles: Synthesis and Application for Nanomedicine," *Int J Mol Sci*, vol. 20, no. 4, p. 865, Feb. 2019, doi: 10.3390/ijms20040865.

[15] "ITIS - Report: Herniaria hirsuta var. hirsuta. Accessed: Jul. 17, 2023. [Online]. Available: https://www.itis.gov/servlet/SingleRpt/SingleRpt?search_topic=TSN&search_value=823679#n ull

[16] "USDA-ARS Germplasm Resources Information Network (GRIN)." Accessed: Jul. 17, 2023. [Online]. Available: https://www.ars-grin.gov/

[17] K. Ammor, D. Bousta, S. Jennan, B. Bennani, A. Chaqroune, and F. Mahjoubi, "Phytochemical Screening, Polyphenols Content, Antioxidant Power, and Antibacterial Activity of Herniaria hirsuta from Morocco," *ScientificWorldJournal*, vol. 2018, p. 7470384, Oct. 2018, doi: 10.1155/2018/7470384.

[18] L. Peeters *et al*, "Compound Characterization and Metabolic Profile Elucidation after In Vitro Gastrointestinal and Hepatic Biotransformation of an Herniaria hirsuta Extract Using Unbiased Dynamic Metabolomic Data Analysis," *Metabolites*, vol. 10, no. 3, p. 111, Mar. 2020, doi: 10.3390/metabo10030111.

[19] "USDA-ARS Germplasm Resources Information Network (GRIN)." Accessed: Jul. 17, 2023. [Online]. Available: https://www.ars-grin.gov/

[20] Y. Bahammou *et al*, "Water sorption isotherms and drying characteristics of rupturewort (Herniaria hirsuta) during a convective solar drying for a better conservation," *Solar Energy*, vol. 201, pp. 916-926, May 2020, doi: 10.1016/j.solener.2020.03.071.

[21] C. L. Keen, R. R. Holt, P. I. Oteiza, C. G. Fraga, and H. H. Schmitz, "Cocoa antioxidants and cardiovascular health2, 3," *The American Journal of Clinical Nutrition*, vol. 81, no. 1, pp. 298S- 303S, Jan. 2005, doi: 10.1093/ajcn/81.1.298S.

[22] I. Van Dooren *et al*, "Cholesterol lowering effect in the gall bladder of dogs by a standardized infusion of Herniaria hirsuta L.," *Journal of Ethnopharmacology*, vol. 169, pp. 69-75, Jul. 2015, doi: 10.1016/j.jep.2015.03.081.

[23] "Search: species: Hemiaria hirsuta AND (dynamicProperties_diffusionGP: "true") | Search Result | SINP - Patrinat." Accessed: Jul. 16, 2023. [Online]. Available : https://openobs.mnhn.fr/openobs-hub/occurrences/search?q=%28dynamicProperties_diffusionGP%3A%22true%22%29&taxa=1 01412#tab_mapView

[24] K. Ammor, D. Bousta, S. Jennan, B. Bennani, A. Chaqroune, and F. Mahjoubi, "Phytochemical Screening, Polyphenols Content, Antioxidant Power, and Antibacterial Activity of Herniaria hirsuta from Morocco," *ScientificWorldJournal*, vol. 2018, p. 7470384, Oct. 2018, doi: 10.1155/2018/7470384.

[25] A. Khouchlaa, M. Tijane, A. Chebat, S. Hseini, and A. Kahouadji, "Enquete ethnopharmacologique des plantes utilises dans le traitement de la lithiase urinaire au Maroc," *Phytotherapie*, vol. 15, no. 5, pp. 274-287, Oct. 2017, doi: 10.1007/s10298-016-1073-4.

[26] F. Atmani, Y. Slimani, M. Mimouni, M. Aziz, B. Hacht, and A. Ziyyat, "Effect of aqueous extract from Herniaria hirsuta L. on experimentally nephrolithiasic rats," *Journal of Ethnopharmacology*, vol. 95, no. 1, pp. 87-93, Nov. 2004, doi: 10.1016/j.jep.2004.06.028.

[27] H. Mabrouki, C. M. M. Duarte, and D. E. Akretche, "Estimation of Total Phenolic Contents and In Vitro Antioxidant and Antimicrobial Activities of Various Solvent Extracts of Melissa officinalis L.," *Arab J Sci Eng*, vol. 43, no. 7, pp. 3349-3357, Jul. 2018, doi: 10.1007/s13369- 017-3000-6.

[28] F. Meiouet, S. El Kabbaj, and M. Daudon, "Etude in vitro de l'activite litholytique de quatre plantes medicinales vis-a-vis des calculs urinaires de cystine," *Progres en Urologie*, vol. 21, no. 1, pp. 40-47, Jan. 2011, doi: 10.1016/j.purol.2010.05.009.

[29] H. Mabrouki, C. M. M. Duarte, and D. E. Akretche, "Estimation of Total Phenolic Contents and In Vitro Antioxidant and Antimicrobial Activities of Various Solvent Extracts of Melissa officinalis L.," *Arab J Sci Eng*, vol. 43, no. 7, pp. 3349-3357, Jul. 2018, doi: 10.1007/s13369- 017-3000-6.

[30] S. Achour *et al*, "Ethnobotanical Study of Medicinal Plants Used as Therapeutic Agents to Manage Diseases of Humans," *Evidence-Based Complementary and Alternative Medicine*, vol. 2022, p. e4104772, Feb. 2022, doi: 10.1155/2022/4104772.

[31] J. Kolodziejczyk-Czepas, S. Kozachok, L. Pecio, S. Marchyshyn, and W. Oleszek, "Determination of phenolic profiles of *Herniaria polygama* and *Herniaria incana* fractions and their *in vitro* antioxidant and anti-inflammatory effects," *Phytochemistry*, vol. 190, p. 112861, Oct. 2021, doi: 10.1016/j.phytochem.2021.112861.

[32] S. Bayda, M. Adeel, T. Tuccinardi, M. Cordani, and F. Rizzolio, "The History of Nanoscience and Nanotechnology: From Chemical-Physical Applications to Nanomedicine, "*Molecules,* vol. 25, no. 1, p. 112, Dec. 2019, doi: 10.3390/molecules25010112.

[33] D. Schaming and H. Remita, "Nanotechnology: from the ancient time to nowadays," *Foundations of Chemistry*, vol. 17, pp. 187-205, Jul. 2015, doi: 10.1007/s10698-015-9235-y.

[34] Afsset, "Securite au travail" Les *nanomateriaux,* Saisine Afsset n° 2006/006, 11 July 2008.

[35] C. Noirot, L. Cormier, N. Schibille, N. Menguy, N. Trcera, and E. Fonda, "Comparative Investigation of Red and Orange Roman Tesserae: Role of Cu and Pb in Colour Formation," *Heritage,* vol. 5, no. 3, pp. 2628-2645, Sep. 2022, doi: 10.3390/heritage5030137.

[36] F. Montanarella and M. V. Kovalenko, "Three Millennia of Nanocrystals," *ACS Nano,* vol. 16, no. 4, pp. 5085-5102, Apr. 2022, doi: 10.1021/acsnano.1c11159.

[37] M. Tahir *et al,* "Cuprous Oxide Nanoparticles: Synthesis, Characterization, and Their Application for Enhancing the Humidity-Sensing Properties of Poly(dioctylfluorene)," *Polymers,* vol. 14, no. 8, p. 1503, Apr. 2022, doi: 10.3390/polym14081503.

[38] E. O. Mikhailova, "Gold Nanoparticles: Biosynthesis and Potential of Biomedical Application," *J Funct Biomater,* vol. 12, no. 4, p. 70, Dec. 2021, doi: 10.3390/jfb12040070.

[39] L. A. Dykman and N. G. Khlebtsov, "Gold Nanoparticles in Biology and Medicine: Recent Advances and Prospects," *Acta Naturae,* vol. 3, no. 2, pp. 34-55, 2011.

[40] D. Thompson, "Michael Faraday's recognition of ruby gold: the birth of modern nanotechnology," *Gold Bulletin,* vol. 40, no. 4, p. 267, 2007.

[41] S. Malik, K. Muhammad, and Y. Waheed, "Nanotechnology: A Revolution in Modern Industry," *Molecules,* vol. 28, no. 2, p. 661, Jan. 2023, doi: 10.3390/molecules28020661.

[42] K. Vijayaraghavan and T. Ashokkumar, "Plant-mediated biosynthesis of metallic nanoparticles: A review of literature, factors affecting synthesis, characterization techniques and applications," *Journal of Environmental Chemical Engineering,* vol. 5, no. 5, pp. 4866-4883, Oct. 2017, doi: 10.1016/j.jece.2017.09.026.

[43] G. Tegart, "Nanotechnology: The Technology for the 21st Century," 2003.

[44] S. Ramanathan, S. C. B. Gopinath, M. K. M. Arshad, P. Poopalan, and V. Perumal, "2 - Nanoparticle synthetic methods: strength and limitations," in *Nanoparticles in Analytical and Medical Devices,* S. C. B. Gopinath and F. Gang, Eds, Elsevier, 2021, pp. 31-43. doi: 10.1016/B978-0-12-821163-2.00002-9.

[45] N. Baig, I. Kammakakam, and W. Falath, "Nanomaterials: a review of synthesis methods, properties, recent progress, and challenges," *Mater. Adv.* vol. 2, no. 6, pp. 1821-1871, Mar. 2021, doi: 10.1039/D0MA00807A.

[46] B. Mekuye and B. Abera, "Nanomaterials: An overview of synthesis, classification, characterization, and applications," *Nano Select,* vol. 4, no. 8, pp. 486-501, 2023, doi: 10.1002/nano.202300038.

[47] M. Goutayer, "Nano-emulsions pour la vectorisation d'agents therapeutiques ou diagnostiques : etude de la biodistribution par imagerie de fluorescence in vivo," phdthesis, Universite Pierre et Marie Curie - Paris VI, 2008.

[48] S. Iravani, "Green synthesis of metal nanoparticles using plants," *Green Chem.* vol. 13, no. 10, pp. 2638-2650, Jan. 2011, doi: 10.1039/C1GC15386B.

[49] H.-Y. Chuang and D.-H. Chen, "Fabrication and photoelectrochemical study of Ag@TiO2 nanoparticle thin film electrode," *International Journal of Hydrogen Energy,* vol. 36, no. 16, pp. 9487-9495, Aug. 2011, doi: 10.1016/j.ijhydene.2011.05.093.

[50] R. Patel (Kumar), P. Bobde, V. Singh (K.), D. Panchal, and S. Pal, "Chapter 21 - Synthesis and applications of carbon nanomaterials-based sensors," in *Advanced Nanomaterials for Point of Care Diagnosis and Therapy,* S. Dave, J. Das, and S. Ghosh, Eds., Elsevier, 2022, pp. 451-476. doi: 10.1016/B978-0-323-85725-3.00019-2.

[51] S. Griffin *et al,* "Natural Nanoparticles: A Particular Matter Inspired by Nature," *Antioxidants (Basel),* vol. 7, no. 1, p. 3, Dec. 2017, doi: 10.3390/antiox7010003.

[52] A. Barhoum *et al,* "Review on Natural, Incidental, Bioinspired, and Engineered Nanomaterials: History, Definitions, Classifications, Synthesis, Properties, Market, Toxicities, Risks, and Regulations," *Nanomaterials (Basel),* vol. 12, no. 2, p. 177, Jan. 2022, doi: 10.3390/nano12020177.

[53] I. Khan, K. Saeed, and I. Khan, "Nanoparticles: Properties, applications and toxicities," *Arabian*

Journal of Chemistry, vol. 12, no. 7, pp. 908-931, Nov. 2019, doi: 10.1016/j.arabjc.2017.05.011.

[54] R. Aswani and E. K. Radhakrishnan, "Chapter 5 - Green approaches for nanotechnology," in *Green Functionalized Nanomaterials for Environmental Applications*, U. Shanker, C. M. Hussain, and M. Rani, Eds. in Micro and Nano Technologies. Elsevier, 2022, pp. 129-154. doi: 10.1016/B978-0-12-823137-1.00005-1.

[55] E. Abbasi *et al*, "Dendrimers: synthesis, applications, and properties," *Nanoscale Res Lett*, vol. 9, no. 1, p. 247, May 2014, doi: 10.1186/1556-276X-9-247.

[56] G. Wu *et al*, "Recent advancement of bioinspired nanomaterials and their applications: A review," *Front Bioeng Biotechnol*, vol. 10, p. 952523, Sep. 2022, doi: 10.3389/fbioe.2022.952523.

[57] P. Slepicka, N. Slepickova Kasalkova, J. Siegel, Z. Kolska, and V. Svorak, "Methods of Gold and Silver Nanoparticles Preparation," *Materials*, vol. 13, no. 1, Art. no. 1, Jan. 2020, doi: 10.3390/ma13010001.

[58] C. Buzea and I. Pacheco, "Chapter 2 - Gold and silver nanoparticles: Properties and toxicity," in *Gold and Silver Nanoparticles*, S. Sahoo and M. R. Hormozi-Nezhad, Eds. in Micro and Nano Technologies. Elsevier, 2023, pp. 59-82. doi: 10.1016/B978-0-323-99454-5.00007-X.

[59] L. R. Adil *et al*, "Chapter 7 - Nanomaterials for sensors: Synthesis and applications," in *Advanced Nanomaterials for Point of Care Diagnosis and Therapy*, S. Dave, J. Das, and S. Ghosh, Eds, Elsevier, 2022, pp. 121-168. doi: 10.1016/B978-0-323-85725-3.00017-9.

[60] A. Klinkova and H. Therien-Aubin, "Chapter 3 - Inorganic nanoparticles," in *Nanochemistry*, A. Klinkova and H. Therien-Aubin, Eds., Elsevier, 2024, pp. 49-110. doi: 10.1016/B978-0-443- 21447-9.00001-1.

[61] K. C. Majhi and M. Yadav, "Chapter 5 - Synthesis of inorganic nanomaterials using carbohydrates," in *Green Sustainable Process for Chemical and Environmental Engineering and Science*, Inamuddin, R. Boddula, M. I. Ahamed, and A. M. Asiri, Eds, Elsevier, 2021, pp. 109135. doi: 10.1016/B978-0-12-821887-7.00003-3.

[62] M. G. Berhe and Y. T. Gebreslassie, "Biomedical Applications of Biosynthesized Nickel Oxide Nanoparticles," *Int J Nanomedicine,* vol. 18, pp. 4229-4251, Jul. 2023, doi: 10.2147/IJN.S410668.

[63] D. Ziental *et al*, "Titanium Dioxide Nanoparticles: Prospects and Applications in Medicine," *Nanomaterials (Basel)*, vol. 10, no. 2, p. 387, Feb. 2020, doi: 10.3390/nano10020387.

[64] G. Maduraiveeran and W. Jin, "Carbon nanomaterials: Synthesis, properties and applications in electrochemical sensors and energy conversion systems," *Materials Science and Engineering : B*, vol. 272, p. 115341, Oct. 2021, doi: 10.1016/j.mseb.2021.115341.

[65] K. D. Patel, R. K. Singh, and H.-W. Kim, "Carbon-based nanomaterials as an emerging platform for theranostics," *Mater. Horiz.* vol. 6, no. 3, pp. 434-469, Mar. 2019, doi: 10.1039/C8MH00966J.

[66] N. Rao, R. Singh, and L. Bashambu, "Carbon-based nanomaterials: Synthesis and prospective applications," *Materials Today: Proceedings*, vol. 44, pp. 608-614, Jan. 2021, doi: 10.1016/j.matpr.2020.10.593.

[67] F. Ahmad *et al*, "Unique Properties of Surface-Functionalized Nanoparticles for BioApplication: Functionalization Mechanisms and Importance in Application," *Nanomaterials (Basel)*, vol. 12, no. 8, p. 1333, Apr. 2022, doi: 10.3390/nano12081333.

[68] A. Oake, P. Bhatt, and Y. V. Pathak, "Understanding Surface Characteristics of Nanoparticles," in *Surface Modification of Nanoparticles for Targeted Drug Delivery*, Y. V. Pathak, Ed., Cham: Springer International Publishing, 2019, pp. 1-17. doi: 10.1007/978-3-030-06115-9_1.

[69] P. Slepicka, N. Slepickova Kasalkova, J. Siegel, Z. Kolska, and V. Svorak, "Methods of Gold and Silver Nanoparticles Preparation," *Materials (Basel)*, vol. 13, no. 1, p. 1, Dec. 2019, doi: 10.3390/ma13010001.

[70] X. Chen, Y. Liu, B. Wang, X. Liu, and C. Lu, "Understanding role of microstructures of nanomaterials in electrochemiluminescence properties and their applications," *TrAC Trends in Analytical Chemistry*, vol. 162, p. 117030, May 2023, doi: 10.1016/j.trac.2023.117030.

[71] H. Kang *et al*, "Stabilization of Silver and Gold Nanoparticles: Preservation and Improvement of Plasmonic Functionalities," *Chem. Rev.* vol. 119, no. 1, pp. 664-699, Jan. 2019, doi: 10.1021/acs.chemrev.8b00341.

[72] R. Schurmann *et al*, "The electronic structure of the metal-organic interface of isolated ligand coated gold nanoparticles," *Nanoscale Adv.* vol. 4, no. 6, pp. 1599-1607, Mar. 2022, doi: 10.1039/D1NA00737H.

[73] Q. Wu, W. Miao, Y. Zhang, H. Gao, and D. Hui, "Mechanical properties of nanomaterials: A review," *Nanotechnology Reviews*, vol. 9, no. 1, pp. 259-273, Jan. 2020, doi: 10.1515/ntrev- 2020-0021.

[74] D. Guo, G. Xie, and J. Luo, "Mechanical properties of nanoparticles: basics and applications," *J. Phys. D: Appl. Phys.* vol. 47, no. 1, p. 013001, Dec. 2013, doi: 10.1088/0022-3727/47/1/013001.

[75] S. Sajjad, S. A. K. Leghari, N.-U.-A. Ryma, and S. A. Farooqi, "Green Synthesis of Metal-Based Nanoparticles and Their Applications," in *Green Metal Nanoparticles*, John Wiley & Sons, Ltd, 2018, pp. 23-77. doi: 10.1002/9781119418900.ch2.

[76] A. V. Rane, K. Kanny, V. K. Abitha, and S. Thomas, "Chapter 5 - Methods for Synthesis of Nanoparticles and Fabrication of Nanocomposites," in *Synthesis of Inorganic Nanomaterials*, S. Mohan Bhagyaraj, O. S. Oluwafemi, N. Kalarikkal, and S. Thomas, Eds. in Micro and Nano Technologies. Woodhead Publishing, 2018, pp. 121-139. doi: 10.1016/B978-0-08-101975- 7.00005-1.

[77] S. Kulkarni, "Synthesis of Nanomaterials-I (Physical Methods)," 2015, pp. 55-76. doi: 10.1007/978-3-319-09171-6_3.

[78] P. U. Ingle, A. P. Ingle, R. R. Philippini, and S. S. da Silva, "4 - Emerging role of nanotechnology in precision farming," in *Nanotechnology in Agriculture and Agroecosystems*, A. P. Ingle, Ed. in Micro and Nano Technologies. Elsevier, 2023, pp. 71-91. doi: 10.1016/B978-0-323-99446- 0.00007-6.

[79] Afsset, "Effets sur la sante de 1 homme et sur l'environnement," (Effects on human health and the environment) *Les nanomateriaux,* n° 2005/010, July 2006.

[80] D. Bokov *et al*, "Nanomaterial by Sol-Gel Method: Synthesis and Application," *Advances in Materials Science and Engineering*, vol. 2021, p. e5102014, Dec. 2021, doi: 10.1155/2021/5102014.

[81] M. Paulose, G. K. Mor, O. K. Varghese, K. Shankar, and C. A. Grimes, "Visible light photoelectrochemical and water-photoelectrolysis properties of titania nanotube arrays," *Journal of Photochemistry and Photobiology A: Chemistry*, vol. 178, no. 1, pp. 8-15, Feb. 2006, doi: 10.1016/j.jphotochem.2005.06.013.

[82] K. Hachem *et al*, "Methods of Chemical Synthesis in the Synthesis of Nanomaterial and Nanoparticles by the Chemical Deposition Method: A Review," *BioNanoSci*, vol. 12, no. 3, pp. 1032-1057, Sep. 2022, doi: 10.1007/s12668-022-00996-w.

[83] N. H. Nam and N. H. Luong, "Nanoparticles: synthesis and applications," *Materials for Biomedical Engineering*, pp. 211-240, 2019, doi: 10.1016/B978-0-08-102814-8.00008-1.

[84] O. Messaoudi, M. Bendahou, O. Messaoudi, and M. Bendahou, "Biological Synthesis of Nanoparticles Using Endophytic Microorganisms: Current Development," in *Nanotechnology and the Environment*, IntechOpen, 2020. doi: 10.5772/intechopen.93734.

[85] M. Boholm, "The use and meaning of nano in American English: Towards a systematic description," *Ampersand*, vol. 3, pp. 163-173, Jan. 2016, doi: 10.1016/j.amper.2016.10.001.

[86] X. Zhu *et al*, "Chitosan-based nanoparticle co-delivery of docetaxel and curcumin ameliorates anti-tumor chemoimmunotherapy in lung cancer," *Carbohydrate Polymers*, vol. 268, p. 118237, Sep. 2021, doi: 10.1016/j.carbpol.2021.118237.

[87] Y. Zhang, F. Fang, L. Li, and J. Zhang, "Self-Assembled Organic Nanomaterials for Drug Delivery, Bioimaging, and Cancer Therapy," *ACS Biomater. Sci. Eng.* vol. 6, no. 9, pp. 48164833, Sep. 2020, doi: 10.1021/acsbiomaterials.0c00883.

[88] T. Liu *et al*, "Silver nanoparticle-functionalized 3D flower-like copper (II)-porphyrin framework nanocomposites as signal enhancers for fabricating a sensitive glutathione electrochemical sensor,"

Sensors and Actuators B: Chemical, vol. 342, p. 130047, Sep. 2021, doi: 10.1016/j.snb.2021.130047.

[89] Y. Wang, J. Xu, L. Shi, and H. Yang, "Recent advances in the antilung cancer activity of biosynthesized gold nanoparticles," *Journal of Cellular Physiology*, vol. 235, no. 12, pp. 89518957, 2020, doi: 10.1002/jcp.29789.

[90] D. Li *et al*, "A Review on Scaling Up Perovskite Solar Cells," *Advanced Functional Materials*, vol. 31, no. 12, pp. 2008621, 2021, doi: 10.1002/adfm.202008621.

[91] H. Zheng *et al*, "Recent developments and challenges of Li-rich Mn-based cathode materials for high-energy lithium-ion batteries," *Materials Today Energy*, vol. 18, p. 100518, Dec. 2020, doi: 10.1016/j.mtener.2020.100518.

[92] Z. Zhang, C. Zhang, H. Zheng, and H. Xu, "Plasmon-Driven Catalysis on Molecules and Nanomaterials," *Acc. Chem. Res.* vol. 52, no. 9, pp. 2506-2515, Sep. 2019, doi: 10.1021/acs.accounts.9b00224.

[93] Q. Li *et al*, "Perspective on theoretical methods and modeling relating to electro-catalysis processes," *Chem. Commun.* vol. 56, no. 69, pp. 9937-9949, Aug. 2020, doi: 10.1039/D0CC02998J.

[94] Z. Wang, F. Gauvin, P. Feng, H. J. H. Brouwers, and Q. Yu, "Self-cleaning and air purification performance of Portland cement paste with low dosages of nanodispersed TiO_2 coatings," *Construction and Building Materials*, vol. 263, p. 120558, Dec. 2020, doi: 10.1016/j.conbuildmat.2020.120558.

[95] Q. Chen, M. Zhou, Y. Pan, and Y. Zhang, "Ligand-Enhanced Zero-Valent Iron for Organic Contaminants Degradation: A Mini Review," *Processes*, vol. 11, no. 2, Art. no. 2, Feb. 2023, doi: 10.3390/pr11020620.

[96] H. Onyeaka, P. Passaretti, T. Miri, and Z. T. Al-Sharify, "The safety of nanomaterials in food production and packaging," *Current Research in Food Science*, vol. 5, pp. 763-774, Jan. 2022, doi: 10.1016/j.crfs.2022.04.005.

[97] Z. Alhalili, "Metal Oxides Nanoparticles: General Structural Description, Chemical, Physical, and Biological Synthesis Methods, Role in Pesticides and Heavy Metal Removal through Wastewater Treatment," *Molecules*, vol. 28, no. 7, p. 3086, Mar. 2023, doi: 10.3390/molecules28073086.

[98] X. Li, J. Natsuki, and T. Natsuki, "Silver nanoparticles/graphene oxide nanoscroll composites synthesized by one step," *Physica E: Low-dimensional Systems and Nanostructures*, vol. 124, p. 114249, Oct. 2020, doi: 10.1016/j.physe.2020.114249.

[99] P. S. Nayak *et al*, "Silver nanoparticles fabricated using medicinal plant extracts show enhanced antimicrobial and selective cytotoxic propensities," *IET Nanobiotechnology*, vol. 13, no. 2, pp. 193-201, 2019, doi: 10.1049/iet-nbt.2018.5025.

[100] M. Ma, B. J. Trzesniewski, J. Xie, and W. A. Smith, "Selective and Efficient Reduction of Carbon Dioxide to Carbon Monoxide on Oxide-Derived Nanostructured Silver Electrocatalysts," *Angewandte Chemie*, vol. 128, no. 33, pp. 9900-9904, 2016, doi: 10.1002/ange.201604654.

[101] M. S. S. Danish, L. L. Estrella-Pajulas, I. M. Alemaida, M. L. Grilli, A. Mikhaylov, and T. Senjyu, "Green Synthesis of Silver Oxide Nanoparticles for Photocatalytic Environmental Remediation and Biomedical Applications," *Metals*, vol. 12, no. 5, Art. no. 5, May 2022, doi: 10.3390/met12050769.

[102] H. H. Nayel and H. S. AL-Jumaili, "Synthesis and characterization of silver oxide nanoparticles prepared by chemical bath deposition for NH3 gas sensing applications," *eijs*, pp. 772-779, Apr. 2020, doi: 10.24996/ijs.2020.61.4.9.

[103] B. N. Rashmi *et al*, "Facile green synthesis of silver oxide nanoparticles and their electrochemical, photocatalytic and biological studies," *Inorganic Chemistry Communications*, vol. 111, p. 107580, Jan. 2020, doi: 10.1016/j.inoche.2019.107580.

[104] M. Cobos, I. De-La-Pinta, G. Quindos, M. J. Fernandez, and M. D. Fernandez, "Synthesis, Physical, Mechanical and Antibacterial Properties of Nanocomposites Based on Poly(vinyl alcohol)/Graphene Oxide-Silver Nanoparticles," *Polymers*, vol. 12, no. 3, Art. no. 3, Mar. 2020, doi: 10.3390/polym12030723.

[105]   H. Daoudi *et al*, "Secondary Metabolite from Nigella Sativa Seeds Mediated Synthesis of Silver Oxide Nanoparticles for Efficient Antioxidant and Antibacterial Activity," *J Inorg Organomet Polym*, vol. 32, no. 11, pp. 4223-4236, Nov. 2022, doi: 10.1007/s10904-022-02393-y.

[106]   N. R. Kokila *et al*, "Thunbergia mysorensis mediated nano silver oxide for enhanced antibacterial, antioxidant, anticancer potential and in vitro hemolysis evaluation," *Journal of Molecular Structure*, vol. 1255, p. 132455, May 2022, doi: 10.1016/j.molstruc.2022.132455.

[107]   S. Z. Hasan, K. N. Ahmad, W. N. R. W. Isahak, M. S. Masdar, and J. M. Jahim, "Synthesis of low-cost catalyst NiO(111) for CO2 hydrogenation into short-chain carboxylic acids," *International Journal of Hydrogen Energy,* vol. 45, no. 42, pp. 22281-22290, Aug. 2020, doi: 10.1016/j.ijhydene.2019.09.102.

[108]   S. Chen *et al*, "High Performance Flexible Lithium-Ion Battery Electrodes: Ion Exchange Assisted Fabrication of Carbon Coated Nickel Oxide Nanosheet Arrays on Carbon Cloth," *Advanced Functional Materials*, vol. 31, no. 24, pp. 2101199, 2021, doi: 10.1002/adfm.202101199.

[109]   S. Yousaf et al, "Tuning the structural, optical and electrical properties of NiO nanoparticles prepared by wet chemical route," *Ceramics International*, vol. 46, no. 3, pp. 3750-3758, Feb. 2020, doi: 10.1016/j.ceramint.2019.10.097.

[110]   F.-R. Juang, C.-H. Hsieh, I.-Y. Huang, W.-Y. Wang, W.-B. Lin, and L. Yen, "Dispersed and spherically assembled porous NiO nanosheets for low concentration ammonia gas sensing applications," *Solid-State   Electronics*, vol. 189, pp. 108224, Mar. 2022, doi: 10.1016/j.sse.2021.108224.

[111]   P. Stremoukhov, D. C. S, A. Safin, S. Nikitov, and A. Kirilyuk, "Phononic manipulation of antiferromagnetic domains in NiO," *New J. Phys.* vol. 24, no. 2, p. 023009, Feb. 2022, doi: 10.1088/1367-2630/ac4ce4.

[112]   N. A. Khan *et al*, "Efficient photodegradation of orange II dye by nickel oxide nanoparticles and nanoclay supported nickel oxide nanocomposite," *Appl Water Sci*, vol. 12, no. 6, p. 131, Apr. 2022, doi: 10.1007/s13201-022-01647-x.

[113]   K. Parvathi and M. T. Ramesan, "Structure, properties, and antibacterial behavior of nickel oxide reinforced natural rubber nanocomposites for flexible electronic applications," *Journal of Applied Polymer Science*, vol. 139, no. 45, p. e53120, 2022, doi: 10.1002/app.53120.

[114]   T. Riaz *et al*, "Phyto-mediated synthesis of nickel oxide (NiO) nanoparticles using leaves' extract of Syzygium cumini for antioxidant and dyes removal studies from wastewater," *Inorganic Chemistry Communications*, vol. 142, p. 109656, Aug. 2022, doi: 10.1016/j.inoche.2022.109656.

[115]   R. Gobi and R. S. Babu, "In-vitro study on chitosan/PVA incorporated with nickel oxide nanoparticles for wound healing application," *Materials Today Communications*, vol. 34, p. 105154, Mar. 2023, doi: 10.1016/j.mtcomm.2022.105154.

[116]   K. Arshak, O. Korostynska, and F. Fahim, "Various Structures Based on Nickel Oxide Thick Films as Gamma Radiation Sensors," *Sensors*, vol. 3, no. 6, Art. no. 6, Jun. 2003, doi: 10.3390/s30600176.

[117]   B. Lavina, P. Dera, and R. T. Downs, "Modern X-ray Diffraction Methods in Mineralogy and Geosciences," *Reviews in Mineralogy and Geochemistry*, vol. 78, no. 1, pp. 1-31, Jan. 2014, doi: 10.2138/rmg.2014.78.1.

[118]   G. Boveri, A. Corozzi, F. Veronesi, and M. Raimondo, "Different Approaches to Low-Wettable Materials for Freezing Environments: Design, Performance and Durability," *Coatings*, vol. 11, p. 77, Jan. 2021, doi: 10.3390/coatings11010077.

[119]   S. Srivastava and A. Bhargava, *Green Nanoparticles: The Future of Nanobiotechnology*. Singapore: Springer Singapore, 2022. doi: 10.1007/978-981-16-7106-7.

[120]   S. Laouini, A. Bouafia, and M. Tedjani, "Catalytic Activity for Dye Degradation and Characterization of Silver/Silver Oxide Nanoparticles Green Synthesized by Aqueous Leaves Extract of Phoenix Dactylifera L.," In Review, preprint, Jan. 2021. doi: 10.21203/rs.3.rs- 139856/v1.

[121]   G. Chuto and P. Chaumet-Riffaud, "Les nanoparticules, "*Medecine Nucleaire,* vol. 34, no. 6,

pp. 370-376, Jun. 2010, doi: 10.1016/j.mednuc.2010.03.003.

[122] S. Shanthi, B. D. Jayaseelan, P. Velusamy, S. Vijayakumar, C. T. Chih, and B. Vaseeharan, "Biosynthesis of silver nanoparticles using a probiotic Bacillus licheniformis Dahb1 and their antibiofilm activity and toxicity effects in Ceriodaphnia cornuta," *Microb Pathog*, vol. 93, pp. 70-77, Apr. 2016, doi: 10.1016/j.micpath.2016.01.014.

[123] A. Lateef, B. I. Folarin, S. M. Oladejo, P. O. Akinola, L. S. Beukes, and E. B. Gueguim-Kana, "Characterization, antimicrobial, antioxidant, and anticoagulant activities of silver nanoparticles synthesized from Petiveria alliacea L. leaf extract," *Prep Biochem Biotechnol*, vol. 48, no. 7, pp. 646-652, 2018, doi: 10.1080/10826068.2018.1479864.

[124] S. Boopathi, S. Gopinath, T. Boopathi, V. Balamurugan, R. Rajeshkumar, and M. Sundararaman, "Characterization and Antimicrobial Properties of Silver and Silver Oxide Nanoparticles Synthesized by Cell-Free Extract of a Mangrove-Associated Pseudomonas aeruginosa M6 Using Two Different Thermal Treatments," *Ind. Eng. Chem. Res.* vol. 51, no. 17, pp. 5976-5985, May 2012, doi: 10.1021/ie3001869.

[125] R. Li, Z. Chen, N. Ren, Y. Wang, Y. Wang, and F. Yu, "Biosynthesis of silver oxide nanoparticles and their photocatalytic and antimicrobial activity evaluation for wound healing applications in nursing care," *Journal of Photochemistry and Photobiology B: Biology*, vol. 199, p. 111593, 2019.

[126] M. Ovais *et al*, "Role of plant phytochemicals and microbial enzymes in biosynthesis of metallic nanoparticles," *Appl Microbiol Biotechnol*, vol. 102, no. 16, pp. 6799-6814, Aug. 2018, doi: 10.1007/s00253-018-9146-7.

[127] Y. Ida *et al*, "Direct Electrodeposition of 1.46 eV Bandgap Silver(I) Oxide Semiconductor Films by Electrogenerated Acid," *Chem. Mater*, vol. 20, no. 4, pp. 1254-1256, Feb. 2008, doi: 10.1021/cm702865r.

[128] T. Xaba, M. J. Moloto, M. Al-Shakban, M. A. Malik, P. O'Brien, and N. Moloto, "The effect of temperature on the growth of Ag2O nanoparticles and thin films from bis(2-hydroxy-1-naphthaldehydato)silver(I) complex by the thermal decomposition of spin-coated films," *Materials Science in Semiconductor Processing*, vol. 71, pp. 109-115, Nov. 2017, doi: 10.1016/j.mssp.2017.07.015.

[129] "S. Torabi, M. J. K. Mansoorkhani, A. Majedi, and S. Motevalli, "REVIEW: Synthesis, Medical And Photocatalyst Applications Of Nano-Ag2O," *Journal of Coordination Chemistry*, vol. 73, no. 13, pp. 1861-1880, Jul. 2020, doi: 10.1080/00958972.2020.1806252.

[130] J. Pan, Y. Sun, Z. Wang, P. Wan, X. Liu, and M. Fan, "Nano silver oxide (AgO) as a super high charge/discharge rate cathode material for rechargeable alkaline batteries," *J. Mater. Chem*, vol. 17, no. 45, pp. 4820-4825, Nov. 2007, doi: 10.1039/B711373K.

[131] J. Tominaga, "The application of silver oxide thin films to plasmon photonic devices," *J. Phys.: Condens. Matter*, vol. 15, no. 25, p. R1101, Jun. 2003, doi: 10.1088/0953-8984/15/25/201.

[132] S. Wren, C. Minelli, Y. Pei, and N. Akhtar, "Evaluation of Particle Size Techniques to Support the Development of Manufacturing Scale Nanoparticles for Application in Pharmaceuticals," *Journal of Pharmaceutical Sciences*, vol. 109, no. 7, pp. 2284-2293, Jul. 2020, doi: 10.1016/j.xphs.2020.04.001.

[133] A. D. Russell and W. B. Hugo, "7 Antimicrobial Activity and Action of Silver," in *Progress in Medicinal Chemistry*, vol. 31, G. P. Ellis and D. K. Luscombe, Eds, Elsevier, 1994, pp. 351370. doi: 10.1016/S0079-6468(08)70024-9.

[134] F. Esmaile, H. Koohestani, and H. Abdollah-Pour, "Characterization and antibacterial activity of silver nanoparticles green synthesized using Ziziphora clinopodioides extract," *Environmental Nanotechnology, Monitoring & Management*, vol. 14, p. 100303, Dec. 2020, doi: 10.1016/j.enmm.2020.100303.

[135] A. Shah, S. Haq, W. Rehman, M. Waseem, S. Shoukat, and M. Rehman, "Photocatalytic and antibacterial activities of paeonia emodi mediated silver oxide nanoparticles," *Mater. Res. Express*,

vol. 6, no. 4, p. 045045, Jan. 2019, doi: 10.1088/2053-1591/aafd42.

[136]   B. N. Rashmi *et al*, "Facile green synthesis of silver oxide nanoparticles and their electrochemical, photocatalytic and biological studies," *Inorganic Chemistry Communications*, vol. 111, p. 107580, Jan. 2020, doi: 10.1016/j.inoche.2019.107580.

[137]   S. Ravichandran, V. Paluri, G. Kumar, K. Loganathan, and B. R. Kokati Venkata, "A novel approach for the biosynthesis of silver oxide nanoparticles using aqueous leaf extract of Callistemon lanceolatus (Myrtaceae) and their therapeutic potential," *Journal of Experimental Nanoscience,* vol. 11, no. 6, pp. 445-458, Apr. 2016, doi: 10.1080/17458080.2015.1077534.

[138]   V. Manikandan *et al*, "Green synthesis of silver oxide nanoparticles and its antibacterial activity against dental pathogens," *3 Biotech*, vol. 7, no. 1, p. 72, Apr. 2017, doi: 10.1007/s13205-017-0670-4.

[139]   S. Menon, R. S., and V. K. S., "A review on biogenic synthesis of gold nanoparticles, characterization, and its applications," *Resource-Efficient Technologies*, vol. 3, no. 4, pp. 516527, Dec. 2017, doi: 10.1016/j.reffit.2017.08.002.

[140]   M. S. Samuel, S. Jose, E. Selvarajan, T. Mathimani, and A. Pugazhendhi, "Biosynthesized silver nanoparticles using Bacillus amyloliquefaciens; Application for cytotoxicity effect on A549 cell line and photocatalytic degradation of p-nitrophenol," *Journal of Photochemistry and Photobiology B: Biology*, vol. 202, p. 111642, Jan. 2020, doi: 10.1016/j.jphotobiol.2019.111642.

[141]   K. Rajendran, V. Karunagaran, B. Mahanty, and S. Sen, "Biosynthesis of hematite nanoparticles and its cytotoxic effect on HepG2 cancer cells," *International Journal of Biological Macromolecules*, vol. 74, pp. 376-381, Mar. 2015, doi: 10.1016/j.ijbiomac.2014.12.028.

[142]   L. Wang, C. Hu, and L. Shao, "The antimicrobial activity of nanoparticles: present situation and prospects for the future," *IJN*, vol. 12, pp. 1227-1249, Feb. 2017, doi: 10.2147/IJN.S121956.

[143]   J. Pan, Y. Sun, Z. Wang, P. Wan, X. Liu, and M. Fan, "Nano silver oxide (AgO) as a super high charge/discharge rate cathode material for rechargeable alkaline batteries," *J. Mater. Chem*, vol. 17, no. 45, pp. 4820-4825, Nov. 2007, doi: 10.1039/B711373K.

[144]   M. Vanaja *et al*, "Degradation of methylene blue using biologically synthesized silver nanoparticles," *Bioinorg Chem Appl*, vol. 2014, pp. 742346, 2014, doi: 10.1155/2014/742346.

[145]   M. I., H. A. M. Saleh, K. M. A. Qasem, M. Shahid, M. Mehtab, and M. Ahmad, "Efficient and selective adsorption and separation of methylene blue (MB) from mixture of dyes in aqueous environment employing a Cu(II) based metal organic framework," *Inorganica Chimica Acta*, vol. 511, p. 119787, Oct. 2020, doi: 10.1016/j.ica.2020.119787.

[146]   A. A. Adeyi, S. N. A. M. Jamil, L. C. Abdullah, T. S. Y. Choong, K. L. Lau, and M. Abdullah, "Adsorptive Removal of Methylene Blue from Aquatic Environments Using Thiourea-Modified Poly(Acrylonitrile-co-Acrylic Acid)," *Materials*, vol. 12, no. 11, Art. no. 11, Jan. 2019, doi: 10.3390/ma12111734.

[147]   K. Ammor, D. Bousta, S. Jennan, B. Bennani, A. Chaqroune, and F. Mahjoubi, "Phytochemical Screening, Polyphenols Content, Antioxidant Power, and Antibacterial Activity of Herniaria hirsuta from Morocco," *ScientificWorldJournal*, vol. 2018, pp. 7470384, 2018, doi: 10.1155/2018/7470384.

[148]   Z. H. Dhoondia and H. Chakraborty, "Lactobacillus Mediated Synthesis of Silver Oxide Nanoparticles," *Nanomaterials and Nanotechnology*, vol. 2, p. 15, Dec. 2012, doi: 10.5772/55741.

[149]   A. K. De, S. Majumdar, S. Pal, S. Kumar, and I. Sinha, "Zn doping induced band gap widening of Ag2O nanoparticles," *Journal of Alloys and Compounds*, vol. 832, p. 154127, Aug. 2020, doi: 10.1016/j.jallcom.2020.154127.

[150]   Y. Liu, P. Li, R. Xue, and X. Fan, "Research on catalytic performance and mechanism of Ag2O/ZnO heterostructure under UV and visible light," *Chemical Physics Letters*, vol. 746, p. 137301, May 2020, doi: 10.1016/j.cplett.2020.137301.

[151]   R. M. Mohamed, Adel. A. Ismail, M. W. Kadi, A. S. Alresheedi, and Ibraheem. A. Mkhalid, "Facile Synthesis of Mesoporous Ag2O-ZnO Heterojunctions for Efficient Promotion of Visible Light

Photodegradation of Tetracycline," *ACS Omega*, vol. 5, no. 51, pp. 33269-33279, Dec. 2020, doi: 10.1021/acsomega.0c04969.

[152] D. Kandi, S. Mansingh, A. Behera, and K. Parida, "Calculation of relative fluorescence quantum yield and Urbach energy of colloidal CdS QDs in various easily accessible solvents," *Journal of Luminescence*, vol. 231, p. 117792, Mar. 2021, doi: 10.1016/j.jlumin.2020.117792.

[153] D. Dharmaraj *et al*, "Antibacterial and cytotoxicity activities of biosynthesized silver oxide (Ag2O) nanoparticles using Bacillus paramycoides," *Journal of Drug Delivery Science and Technology*, vol. 61, p. 102111, Feb. 2021, doi: 10.1016/j.jddst.2020.102111.

[154] S. Mourdikoudis, R. M. Pallares, and N. T. K. Thanh, "Characterization techniques for nanoparticles: comparison and complementarity upon studying nanoparticle properties," *Nanoscale*, vol. 10, no. 27, pp. 12871-12934, Jul. 2018, doi: 10.1039/C8NR02278J.

[155] A. A. Fairuzi, N. N. Bonnia, R. M. Akhir, M. A. Abrani, and H. M. Akil, "Degradation of methylene blue using silver nanoparticles synthesized from imperata cylindrica aqueous extract," *IOP Conf. Ser.: Earth Environ. Sci.* 105, no. 1, p. 012018, Jan. 2018, doi: 10.1088/17551315/105/1/012018.

[156] M. Nasrollahzadeh, Z. Issaabadi, and S. M. Sajadi, "Green synthesis of Cu/Al2O3 nanoparticles as efficient and recyclable catalyst for reduction of 2,4-dinitrophenylhydrazine, Methylene blue and Congo red," *Composites Part B: Engineering*, vol. 166, pp. 112-119, Jun. 2019, doi: 10.1016/j.compositesb.2018.11.113.

[157] S. Raj, H. Singh, R. Trivedi, and V. Soni, "Biogenic synthesis of AgNPs employing Terminalia arjuna leaf extract and its efficacy towards catalytic degradation of organic dyes," *Sci Rep*, vol. 10, no. 1, p. 9616, Jun. 2020, doi: 10.1038/s41598-020-66851-8.

[158] S. D. Khairnar and V. S. Shrivastava, "Facile synthesis of nickel oxide nanoparticles for the degradation of Methylene blue and Rhodamine B dye: a comparative study," *Journal of Taibah University for Science*, vol. 13, no. 1, pp. 1108-1118, Dec. 2019, doi: 10.1080/16583655.2019.1686248.

[159] Z. Cai, Y. Sun, W. Liu, F. Pan, P. Sun, and J. Fu, "An overview of nanomaterials applied for removing dyes from wastewater," *Environ Sci Pollut Res*, vol. 24, no. 19, pp. 15882-15904, Jul. 2017, doi: 10.1007/s11356-017-9003-8.

[160] K. Chaudhary *et al*, "Graphene oxide and reduced graphene oxide supported ZnO nanochips for removal of basic dyes from the industrial effluents," *Fullerenes, Nanotubes and Carbon Nanostructures*, vol. 29, no. 11, pp. 915-928, Nov. 2021, doi: 10.1080/1536383X.2021.1917553.

[161] G. Buyukozkan and F. Gofer, "Digital Supply Chain: Literature review and a proposed framework for future research," *Computers in Industry*, vol. 97, pp. 157-177, May 2018, doi: 10.1016/j.compind.2018.02.010.

[162] Z. M. Alimirzaeva, A. B. Isaev, N. S. Shabanov, A. G. Magomedova, M. V. Kadiev, and K. Kaviyarasu, "Photoelectrocatalytic activity PbO2 loaded highly oriented TiO2 nanotube arrays," *Materials Today: Proceedings*, vol. 36, pp. 325-327, 2021, doi: 10.1016/j.matpr.2020.04.111.

[163] J. Li, Z. Liu, D. Wang, and Z. Zhu, "Visible-light responsive carbon-anatase-hematite core-shell microspheres for methylene blue photodegradation," *Materials Science in Semiconductor Processing*, vol. 27, pp. 950-957, Nov. 2014, doi: 10.1016/j.mssp.2014.08.038.

[164] W. Hu, D. Chu, L. Wang, X. Chen, H. Yang, and J. Sun, "Ultrasound-assisted synthesis of hexagonal cone-like Cu2O architectures with enhanced photocatalytic activity," *Nano-Structures & Nano-Objects*, vol. 12, pp. 220-228, Oct. 2017, doi: 10.1016/j.nanoso.2017.09.018.

[165] J. Singh, T. Dutta, K.-H. Kim, M. Rawat, P. Samddar, and P. Kumar, "'Green' synthesis of metals and their oxide nanoparticles: applications for environmental remediation," *Journal of Nanobiotechnology*, vol. 16, no. 1, p. 84, Oct. 2018, doi: 10.1186/s12951-018-0408-4.

[166] S. F. Ahmed *et al*, "Green approaches in synthesising nanomaterials for environmental nanobioremediation: Technological advancements, applications, benefits and challenges," *Environmental Research*, vol. 204, p. 111967, Mar. 2022, doi: 10.1016/j.envres.2021.111967.

[167]   M. S. Chavali and M. P. Nikolova, "Metal oxide nanoparticles and their applications in nanotechnology," *SN Appl. Sci.* vol. 1, no. 6, p. 607, May 2019, doi: 10.1007/s42452-019-0592- 3.

[168]   C. Kalita and P. Saikia, "Magnetically separable tea leaf mediated nickel oxide nanoparticles for excellent photocatalytic activity," *Journal of the Indian Chemical Society*, vol. 98, no. 11, p. 100213, Nov. 2021, doi: 10.1016/j.jics.2021.100213.

[169]   S. Thirbika, H. Karthi, R. Premila, and M. Ramesh Prabhu, "Investigations on biosynthesized nickel oxide nanoparticles using Cymbopogon citratus leaf extract for antibacterial activity," *Materials Today: Proceedings*, p. S2214785322034496, May 2022, doi: 10.1016/j.matpr.2022.05.168.

[170]   F. Ameen *et al*, "Phytosynthesis of silver nanoparticles using Mangifera indica flower extract as bioreductant and their broad-spectrum antibacterial activity," *Bioorganic Chemistry*, vol. 88, p. 102970, Jul. 2019, doi: 10.1016/j.bioorg.2019.102970.

[171]   A. Shah, S. Haq, W. Rehman, M. Waseem, S. Shoukat, and M. Rehman, "Photocatalytic and antibacterial activities of paeonia emodi mediated silver oxide nanoparticles," *Mater. Res. Express*, vol. 6, no. 4, p. 045045, Jan. 2019, doi: 10.1088/2053-1591/aafd42.

[172]   F. Esmaile, H. Koohestani, and H. Abdollah-Pour, "Characterization and antibacterial activity of silver nanoparticles green synthesized using Ziziphora clinopodioides extract," *Environmental Nanotechnology, Monitoring & Management*, vol. 14, p. 100303, Dec. 2020, doi: 10.1016/j.enmm.2020.100303.

[173]   S. Das, V. K. Singh, A. K. Dwivedy, A. K. Chaudhari, and N. K. Dubey, "Insecticidal and fungicidal efficacy of essential oils and nanoencapsulation approaches for the development of next generation ecofriendly green preservatives for management of stored food commodities: an overview," *International Journal of Pest Management*, vol. 0, no. 0, pp. 1-32, Sep. 2021, doi: 10.1080/09670874.2021.1969473.

[174]   B. El-Ghmari, H. Farah, and A. Ech-Chahad, "A New Approach for the Green Biosynthesis of Silver Oxide Nanoparticles Ag2O, Characterization and Catalytic Application," *Bulletin of Chemical Reaction Engineering & Catalysis*, vol. 16, no. 3, pp. 651-660, Sep. 2021, doi: 10.9767/bcrec.16.3.11577.651-660.

[175]   B. N. Rashmi *et al*, "Facile green synthesis of lanthanum oxide nanoparticles using Centella Asiatica and Tridax plants: Photocatalytic, electrochemical sensor and antimicrobial studies," *Applied Surface Science Advances*,   vol. 7, p. 100210, Feb. 2022, doi:
10.1016/j.apsadv.2022.100210.

[176]   F. Esmaile, H. Koohestani, and H. Abdollah-Pour, "Characterization and antibacterial activity of silver nanoparticles green synthesized using *Ziziphora clinopodioides* extract," *Environmental Nanotechnology, Monitoring & Management*, vol. 14, p. 100303, Dec. 2020, doi: 10.1016/j.enmm.2020.100303.

[177]   S. Menon, R. S., and V. K. S., "A review on biogenic synthesis of gold nanoparticles, characterization, and its applications," *Resource-Efficient Technologies,* vol. 3, no. 4, pp. 516527, Dec. 2017, doi: 10.1016/j.reffit.2017.08.002.

[178]   M. Ovais *et al*, "Role of plant phytochemicals and microbial enzymes in biosynthesis of metallic nanoparticles," *Appl Microbiol Biotechnol*, vol. 102, no. 16, pp. 6799-6814, Aug. 2018, doi: 10.1007/s00253-018-9146-7.

[179]   M. S. Samuel, S. Jose, E. Selvarajan, T. Mathimani, and A. Pugazhendhi, "Biosynthesized silver nanoparticles using Bacillus amyloliquefaciens; Application for cytotoxicity effect on A549 cell line and photocatalytic degradation of p-nitrophenol," *Journal of Photochemistry and Photobiology B: Biology*, vol. 202, p. 111642, Jan. 2020, doi: 10.1016/j.jphotobiol.2019.111642.

[180]   M. Darbandi, M. Eynollahi, N. Badri, M. F. Mohajer, and Z.-A. Li, "NiO nanoparticles with superior sonophotocatalytic performance in organic pollutant degradation," *Journal of Alloys and Compounds*, vol. 889, p. 161706, Dec. 2021, doi: 10.1016/j.jallcom.2021.161706.

[181]   A. Akbari, Z. Sabouri, H. A. Hosseini, A. Hashemzadeh, M. Khatami, and M. Darroudi, "Effect of nickel oxide nanoparticles as a photocatalyst in dyes degradation and evaluation of effective

parameters in their removal from aqueous environments," *Inorganic Chemistry Communications*, vol. 115, p. 107867, May 2020, doi: 10.1016/j.inoche.2020.107867.

[182]   M. Boudiaf *et al*, "Green synthesis of NiO nanoparticles using Nigella sativa extract and their enhanced electro-catalytic activity for the 4-nitrophenol degradation," *Journal of Physics and Chemistry of Solids*, vol. 153, p. 110020, Jun. 2021, doi: 10.1016/j.jpcs.2021.110020.

[183]   P. Bhatia and M. Nath, "Green synthesis of p-NiO/n-ZnO nanocomposites: Excellent adsorbent for removal of congo red and efficient catalyst for reduction of 4-nitrophenol present in wastewater," *Journal of Water Process Engineering*, vol. 33, p. 101017, Feb. 2020, doi: 10.1016/j.jwpe.2019.101017.

[184]   K. Anandan and V. Rajendran, "Effects of Mn on the magnetic and optical properties and photocatalytic activities of NiO nanoparticles synthesized via the simple precipitation process," *Materials Science and Engineering: B*, vol. 199, pp. 48-56, Sep. 2015, doi: 10.1016/j.mseb.2015.04.015.

[185]   T. Adinaveen, T. Karnan, and S. A. Samuel Selvakumar, "Photocatalytic and optical properties of NiO added Nephelium lappaceum L. peel extract: An attempt to convert waste to a valuable product," *Heliyon*, vol. 5, no. 5, p. e01751, May 2019, doi: 10.1016/j.heliyon.2019.e01751.

[186]   S. Farhadi, M. Kazem, and F. Siadatnasab, "NiO nanoparticles prepared via thermal decomposition of the bis(dimethylglyoximato)nickel(II) complex: A novel reusable heterogeneous catalyst for fast and efficient microwave-assisted reduction of nitroarenes with ethanol," *Polyhedron*, vol. 30, no. 4, pp. 606-613, Mar. 2011, doi: 10.1016/j.poly.2010.11.037.

[187]   A. K. Ramasami, M. V. Reddy, and G. R. Balakrishna, "Combustion synthesis and characterization of NiO nanoparticles," *Materials Science in Semiconductor Processing*, vol. 40, pp. 194-202, Dec. 2015, doi: 10.1016/j.mssp.2015.06.017.

[188]   N. N. M. Zorkipli, N. H. M. Kaus, and A. A. Mohamad, "Synthesis of NiO Nanoparticles through Sol-gel Method," *Procedia Chemistry*, vol. 19, pp. 626-631,   2016, doi: 10.1016/j.proche.2016.03.062.

[189]   S. Ghazal *et al*, "Sol-gel biosynthesis of nickel oxide nanoparticles using Cydonia oblonga extract and evaluation of their cytotoxicity and photocatalytic activities," *Journal of Molecular Structure*, vol. 1217, p. 128378, Oct. 2020, doi: 10.1016/j.molstruc.2020.128378.

[190]   M. A. Rahman, R. Radhakrishnan, and R. Gopalakrishnan, "Structural, optical, magnetic and antibacterial properties of Nd doped NiO nanoparticles prepared by co-precipitation method," *Journal of Alloys and Compounds,* vol. 742, pp. 421-429, Apr. 2018, doi: 10.1016/j.jallcom.2018.01.298.

[191]   W.-N. Wang, Y. Itoh, I. W. Lenggoro, and K. Okuyama, "Nickel and nickel oxide nanoparticles prepared from nickel nitrate hexahydrate by a low pressure spray pyrolysis," *Materials Science and Engineering: B*, vol. 111, no. 1, pp. 69-76, Aug. 2004, doi: 10.1016/j.mseb.2004.03.024.

[192]   Wei Zhiqiang, Qiao Hongxia, Yang Hua, Zhang Cairong, and Yan Xiaoyan, "Characterization of NiO nanoparticles by anodic arc plasma method," *Journal of Alloys and Compounds*, vol. 479, no. 1-2, pp. 855-858, Jun. 2009.

[193]   H. G. Gebretinsae, M. G. Tsegay, and Z. Y. Nuru, "Biosynthesis of nickel oxide (NiO) nanoparticles from cactus plant extract," *Materials Today: Proceedings*, vol. 36, pp. 566-570, 2021, doi: 10.1016/j.matpr.2020.05.331.

[194]   Y. Bahammou *et al*, "Water sorption isotherms and drying characteristics of rupturewort (Herniaria hirsuta) during a convective solar drying for a better conservation," *Solar Energy*, vol. 201, pp. 916-926, 2020.

[195]   I. van Dooren *et al*, "Cholesterol lowering effect in the gall bladder of dogs by a standardized infusion of Herniaria hirsuta L.," *Journal of Ethnopharmacology*, vol. 169, pp. 69-75, Jul. 2015, doi: 10.1016/j.jep.2015.03.081.

[196]   Y. Qi, H. Qi, J. Li, and C. Lu, "Synthesis, microstructures and UV-vis absorption properties of 0-Ni(OH)2 nanoplates and NiO nanostructures," *Journal of Crystal Growth*, vol. 310, no. 18, pp. 4221-4225, Aug. 2008, doi: 10.1016/j.jcrysgro.2008.06.047.

[197]   Z. Sabouri, A. Akbari, H. A. Hosseini, M. Khatami, and M. Darroudi, "Green-based bio-synthesis of nickel oxide nanoparticles in Arabic gum and examination of their cytotoxicity, photocatalytic and antibacterial effects," *Green Chemistry Letters and Reviews*, vol. 14, no. 2, pp. 404-414, Apr. 2021, doi: 10.1080/17518253.2021.1923824.

[198]   A. Fall, J. Sackey, N. Mayedwa, and B. D. Ngom, "Investigation of structural and optical properties of CdO nanoparticles via peel of Citrus x sinensis," *Materials Today: Proceedings*, vol. 36, pp. 298-302, 2021, doi: 10.1016/j.matpr.2020.04.057.

[199]   R. Ramesh, V. Yamini, S. J. Sundaram, F. L. A. Khan, and K. Kaviyarasu, "Investigation of structural and optical properties of NiO nanoparticles mediated by Plectranthus amboinicus leaf extract," *Materials Today: Proceedings*, vol. 36, pp. 268-272, 2021, doi: 10.1016/j.matpr.2020.03.581.

[200]   S. Yousaf *et al*, "Tuning the structural, optical and electrical properties of NiO nanoparticles prepared by wet chemical route," *Ceramics International*, vol. 46, no. 3, pp. 3750-3758, Feb. 2020, doi: 10.1016/j.ceramint.2019.10.097.

[201]   M. Ben Amor, A. Boukhachem, K. Boubaker, and M. Amlouk, "Structural, optical and electrical studies on Mg-doped NiO thin films for conductivity applications," *Materials Science in Semiconductor Processing*, vol. 27, pp. 994-1006, Nov. 2014, doi: 10.1016/j.mssp.2014.08.008.

[202]   Z. Sabouri, A. Akbari, H. A. Hosseini, M. Khatami, and M. Darroudi, "Egg white-mediated green synthesis of NiO nanoparticles and study of their cytotoxicity and photocatalytic activity," *Polyhedron*, vol. 178, p. 114351, Mar. 2020, doi: 10.1016/j.poly.2020.114351.

[203]   K. Kannan, D. Radhika, M. P. Nikolova, K. K. Sadasivuni, H. Mahdizadeh, and U. Verma, "Structural studies of bio-mediated NiO nanoparticles for photocatalytic and antibacterial activities," *Inorganic Chemistry Communications*, vol. 113, p. 107755, Mar. 2020, doi: 10.1016/j.inoche.2019.107755.

[204]   K. Karthik, M. Shashank, V. Revathi, and T. Tatarchuk, "Facile microwave-assisted green synthesis of NiO nanoparticles from Andrographis paniculata leaf extract and evaluation of their photocatalytic and anticancer activities," *Molecular Crystals and Liquid Crystals*, vol. 673, no. 1, pp. 70-80, Sep. 2018, doi: 10.1080/15421406.2019.1578495.

[205]   Z. Sabouri, A. Akbari, H. A. Hosseini, A. Hashemzadeh, and M. Darroudi, "Bio-based synthesized NiO nanoparticles and evaluation of their cellular toxicity and wastewater treatment effects," *Journal of Molecular Structure*, vol. 1191, pp. 101-109, Sep. 2019, doi: 10.1016/j.molstruc.2019.04.075.

[206]   Z. Sabouri, A. Akbari, H. A. Hosseini, A. Hashemzadeh, and M. Darroudi, "Eco-Friendly Biosynthesis of Nickel Oxide Nanoparticles Mediated by Okra Plant Extract and Investigation of Their Photocatalytic, Magnetic, Cytotoxicity, and Antibacterial Properties," *J Clust Sci*, vol. 30, no. 6, pp. 1425-1434, Nov. 2019, doi: 10.1007/s10876-019-01584-x.

[207]   Z. Sabouri, A. Akbari, H. A. Hosseini, M. Khatami, and M. Darroudi, "Tragacanth-mediate synthesis of NiO nanosheets for cytotoxicity and photocatalytic degradation of organic dyes," *Bioprocess Biosyst Eng*, vol. 43, no. 7, pp. 1209-1218, Jul. 2020, doi: 10.1007/s00449-020- 02315-7.

[208]   Z. Sabouri, M. Sabouri, M. S. Amiri, M. Khatami, and M. Darroudi, "Plant-based synthesis of cerium oxide nanoparticles using Rheum turkestanicum extract and evaluation of their cytotoxicity and photocatalytic properties," *Materials Technology*, vol. 37, no. 8, pp. 555-568, Jul. 2022, doi: 10.1080/10667857.2020.1863573.

[209]   Z. Sabouri, A. Rangrazi, M. S. Amiri, M. Khatami, and M. Darroudi, "Green synthesis of nickel oxide nanoparticles using Salvia hispanica L. (chia) seeds extract and studies of their photocatalytic activity and cytotoxicity effects," *Bioprocess Biosyst Eng*, vol. 44, no. 11, pp. 2407-2415, Nov. 2021, doi: 10.1007/s00449-021-02613-8.

[210]   A. Miri, F. Mahabbati, A. Najafidoust, M. J. Miri, and M. Sarani, "Nickel oxide nanoparticles: biosynthesized, characterization and photocatalytic application in degradation of methylene blue dye," *Inorganic and Nano-Metal Chemistry*, vol. 52, no. 1, pp. 122-131, Jan. 2022, doi: 10.1080/24701556.2020.1862226.

[211] P. Bhatia and M. Nath, "Nanocomposites of ternary mixed metal oxides (Ag2O/NiO/ZnO) used for the efficient removal of organic pollutants," *Journal of Water Process Engineering*, vol. 49, p. 102961, Oct. 2022, doi: 10.1016/j.jwpe.2022.102961.

[212] M. R. Aravind *et al*, "Influence of Various Concentrations of Cetyltrimethylammonium Bromide on the Properties of Nickel Oxide Nanoparticles for Supercapacitor Application," *NANO*, vol. 16, no. 12, pp. 2150138, Nov. 2021, doi: 10.1142/S1793292021501381.

[213] X. Wan, M. Yuan, S. Tie, and S. Lan, "Effects of catalyst characters on the photocatalytic activity and process of NiO nanoparticles in the degradation of methylene blue," *Applied Surface Science*, vol. 277, pp. 40-46, Jul. 2013, doi: 10.1016/j.apsusc.2013.03.126.

[214] H. Tju, A. Taufik, and R. Saleh, "Enhanced UV Photocatalytic Performance of Magnetic Fe3O4/CuO/ZnO/NGP Nanocomposites," *J. Phys: Conf. Ser.* vol. 710, no. 1, p. 012005, Apr. 2016, doi: 10.1088/1742-6596/710/1/012005.

[215] K. Karthik, S. Dhanuskodi, C. Gobinath, S. Prabukumar, and S. Sivaramakrishnan, "Multifunctional properties of microwave assisted CdO-NiO-ZnO mixed metal oxide nanocomposite: enhanced photocatalytic and antibacterial activities," *J Mater Sci: Mater Electron*, vol. 29, no. 7, pp. 5459-5471, Apr. 2018, doi: 10.1007/s10854-017-8513-y.

[216] N. M. El-Shafai, M. E. El-Khouly, M. El-Kemary, M. S. Ramadan, and M. S. Masoud, "Graphene oxide-metal oxide nanocomposites: fabrication, characterization and removal of cationic rhodamine B dye," *RSC Adv.*, vol. 8, no. 24, pp. 13323-13332, Apr. 2018, doi: 10.1039/C8RA00977E.

[217] T. Munawar, F. Iqbal, S. Yasmeen, K. Mahmood, and A. Hussain, "Multi metal oxide NiO-CdO- ZnO nanocomposite-synthesis, structural, optical, electrical properties and enhanced sunlight driven photocatalytic activity," *Ceramics International*, vol. 46, no. 2, pp. 2421-2437, Feb. 2020, doi: 10.1016/j.ceramint.2019.09.236.

[218] E. F. Aboelfetoh, A. E. Aboubaraka, and E.-Z. M. Ebeid, "Synergistic Effect of Iron and Copper Oxides in the Removal of Organic Dyes Through Thermal Induced Catalytic Degradation Process," *J Clust Sci*, vol. 34, no. 5, pp. 2521-2535, Sep. 2023, doi: 10.1007/s10876-022-02400- 9.

[219] P. Bhatia and M. Nath, "Green synthesis of p-NiO/n-ZnO nanocomposites: Excellent adsorbent for removal of congo red and efficient catalyst for reduction of 4-nitrophenol present in wastewater," *Journal of Water Process Engineering*, vol. 33, p. 101017, Feb. 2020, doi: 10.1016/j.jwpe.2019.101017

Buy your books fast and straightforward online - at one of world's fastest growing online book stores! Environmentally sound due to Print-on-Demand technologies.

Buy your books online at
www.morebooks.shop

Kaufen Sie Ihre Bücher schnell und unkompliziert online – auf einer der am schnellsten wachsenden Buchhandelsplattformen weltweit! Dank Print-On-Demand umwelt- und ressourcenschonend produzi ert.

Bücher schneller online kaufen
www.morebooks.shop

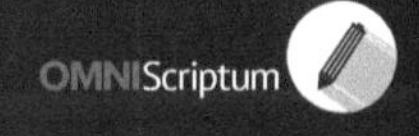

Printed by Books on Demand GmbH, Norderstedt / Germany